T0258768

Surfactants in Emerging Technologies

edited by

Milton J. Rosen

Surfactant Research Institute
Brooklyn College
The City University of New York
Brooklyn, New York

MARCEL DEKKER, INC.   New York and Basel

Library of Congress Cataloging-in-Publication Data

Surfactants in emerging technologies.

    (Surfactant science series ; 26)
    Proceedings of a conference sponsored by the National
Science Foundation, held at Brooklyn College of the City
University of New York, June 6, 1986.
    Includes index.
    1. Surface active agents--Congresses. 2. High
technology--Congresses. I. Rosen, Milton J. II. National
Science Foundation (U.S.) III. Series: Surfactant science
series : v. 26.
TP994.S877 1987   668'.1   87-8968
ISBN 0-8247-7801-4

MARCEL DEKKER, INC.
270 Madison Avenue, New York, New York 10016

Current printing (last digit):
10 9 8 8 7 6 5 4 3 2 1

PRINTED IN THE UNITED STATES OF AMERICA

Preface

On June 6, 1986, a conference on "The Role of Surfactants in New
and Emerging Technology," organized by Professor Milton J. Rosen of
the Department of Chemistry and sponsored by the National Science
Foundation, was held at Brooklyn College of The City University of
New York. About sixty representatives from industry, academia, and
government attended. This volume is based on the proceedings of
that conference.

The conference was held because the past several years had
seen a rapid expansion of surfactant use into a number of new,
important areas of technology. It was therefore felt that the time
had come to examine the position of surfactants in these new growth
areas, the needs for surfactant research to facilitate advances in
those areas, and the resources (or lack thereof) available to meet
those needs.

The program consisted of a morning plenary session at which
invited speakers each discussed surfactant utilization in a particular area of technology. Eight areas were discussed: (1) biotechnology, (2) electronic printing, (3) high-technology electronic
ceramics, (4) magnetic recording, (5) microelectronics, (6) non-

conventional energy production, (7) novel pollution control methods, and (8) novel separation techniques. In the afternoon, there were two 1-hour sessions of group discussions. These sessions consisted of eight simultaneous, small discussion groups, each devoted to one area discussed in the morning session. The morning plenary session's speaker acted as discussion leader with a rapporteur to record the proceedings. Since there were two afternoon sessions, each conference attendee had the opportunity to participate in two group discussions. The findings of each of these group discussions were presented by the rapporteurs at a final plenary session that concluded the conference.

Two constantly recurring themes were apparent in the group duscussions and in conversations with individual conferees: (1) the need for more fundamental research on the mechanisms by which surfactants perform their function in these new technological areas, e.g. charge development at interfaces in solutions of surfactants in hydrocarbons, and (2) the need for surfactants designed specifically for these new applications. One problem is the difficulty, or lack, of communication between surfactant users and surfactant producers. Communication is difficult because of the large amounts of proprietary information involved in the use of surfactants for these purposes. Another problem is the small sales volumes involved, which makes surfactant producers reluctant to undertake the research and development needed to meet application needs.

Milton J. Rosen

Contents

Contributors

LECTURERS

MARK S. CHAGNON Integrated Magnetics Inc., Chelmsford, Massachusetts

MELVIN D. CROUCHER Xerox Research Centre of Canada, Mississauga, Ontario, Canada

ROBERT DONADIO Integrated Magnetics Inc., Chelmsford, Massachusetts

PATRICK G. GRIMES Exxon Research and Engineering Co., Annandale, New Jersey

JEFFREY H. HARWELL Institute for Applied Surfactant Research, and School of Chemical Engineering and Materials Science, University of Oklahoma, Norman, Oklahoma

WILLIAM H. LINDENBERGER NALCO Chemical Co., Naperville, Illinois

SAUL L. NEIDLEMAN Cetus Corporation, Emeryville, California

MELVIN POMERANTZ IBM, T. J. Watson Research Center, Yorktown Heights, New York

JOHN F. SCAMEHORN Institute for Applied Surfactant Research, and School of Chemical Engineering and Materials Science, University of Oklahoma, Norman, Oklahoma

ELLEN S. TORMEY Ceramics Process Systems Corp., Cambridge, Massachusetts

DONALD B. WETLAUFER Department of Chemistry, University of Delaware, Newark, Delaware

RAPPORTEURS

GRAHAM BARKER Witco Chemical Corp., Oakland, New Jersey

CAROLYN A. ERTELL Stauffer Chemical Co., Elmsford, New York

ROBERT FALK CIBA-Geigy Corp., Ardsley, New York

C. HERVE Department of Chemistry, Syracuse University, Syracuse, New York

ROBERT B. LOGIN GAF Corp., Wayne, New Jersey

M. N. MEMERING Pfizer Pigments, Inc., Easton, Pennsylvania

RICHARD M. MULLINS Olin Chemicals Group, Cheshire, Connecticut

DAVID C. NAUGLE Shell Development Co., Houston, Texas

LESLEY NOWAKOWSKI American Cyanamid Co., Stamford, Connecticut

JOHN F. SCAMEHORN Institute for Applied Surface Research, and School of Chemical Engineering and Materials Science, University of Oklahoma, Norman, Oklahoma

STEVEN A. SNOW Dow Corning Corp., Midland, Michigan

DARSH T. WASAN Department of Chemical Engineering, Illinois Institute of Technology, Chicago, Illinois

J. A. WINGRAVE E. I. du Pont de Nemours, Deepwater, New Jersey

DONALD L. WOOD Shell Development Co., Houston, Texas

Surfactants in
Emerging Technologies

1

The Use of Surfactants in Liquid Developers for Electronic Printing

Melvin D. Croucher

Xerox Research Centre of Canada
Mississauga, Ontario, Canada

ABSTRACT

With the advent of microcomputer technology there has been an
increased demand for improved methods of obtaining hardcopy output
of electronically stored documents. Two of the technologies that
have emerged to meet this demand are electrographic and ink-jet
printing. The common feature in these two printing processes is
that they both use liquid developers to produce marks on paper.
However, the methods used by these technologies to convert
electronic signals to hardcopy differ dramatically. In this paper
we briefly review the salient features of both electrographic and
ink-jet printing. This is followed by a description of the liquid
developers that are used in these two printing systems, their
materials design criteria and physico-chemical characteristics.
Finally, we delineate some possible future developer materials
trends with special emphasis being given to the role of surfactants
in these liquid marking materials.

1.0 INTRODUCTION

The placement of electronic document creation and management
systems into the office environment is occurring with increasing
frequency. These devices allow documents to be created,
manipulated and stored in electronic form. This technological
change has resulted in a demand for improved methods of obtaining
hard copy images of these electronic signals, consequently,
numerous printing technologies have emerged to meet this demand.
They include ink-jet, laser, thermal transfer and electrographic
printing as well as the more traditional impact printers.

The major electronic printer options that are presently
available are shown in fig. 1 where the writing energy, which is
the energy required at the print head to generate images, is
plotted against the number of process steps required to generate
the output print. It can be seen that impact printing is a simple
process, and therefore reliable, but requires considerable energy.
Laser printing on the other hand requires little energy to form
images but it is a complex process. This complexity contributes to
its lower reliability. An advantage of these technologies is that
they are capable of printing on plain paper. This lack of
sensitivity to the paper structure occurs because the marking inks
are "dry" which enables thick layers to be placed on the surface of
the paper, so obscuring its texture.

Fig. 1 also shows that the technological advances that are
sought in electronic printing devices are process simplicity,
reliability and a decrease in the energy required to form images.
Of the printing systems presently available, ink-jet and
electrography are the clear leaders in these areas. Both of these
marking devices use liquid toners to develop images on paper,
consequently, the images are sensitive to the paper structure.
Therefore, both technologies require special papers on which to
print. This is usually considered to be a disadvantages for these
printing systems relative to dry marking processes. A major
advantage of these systems however, is their ability to produce

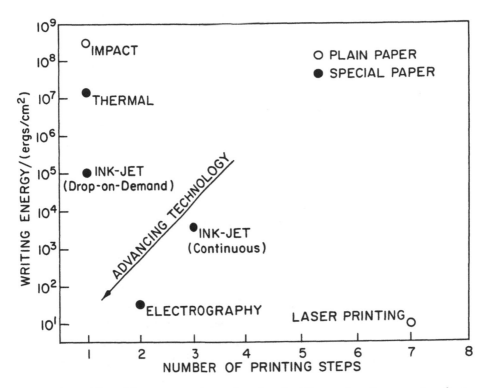

FIG. 1   *The writing energy is shown plotted against the number of process steps for a variety of printing technologies. This has been adapted from A. H. Sporer in "Advances in Non-Impact Printing Technologies for Computer and Office Applications", Ed. J. Gaynor, p. 1338, Van Nostrand Reinhold, 1982.*

color prints and numerous color printing systems based on these technologies are now sold commercially. It is in color printing that liquid marking systems seem to have found their niche.

In this paper we will briefly review the salient features of both electrographic and ink-jet printing. This will be followed by a description of the liquid developers that are used in these printing technologies, their materials design criteria and their physico-chemical and imaging characteristics. Finally, we discuss possible future developer materials trends with special emphasis being given to the role of surfactants in these liquid marking technologies.

2.0   BRIEF REVIEW OF LIQUID PRINTING TECHNOLOGIES

Although both ink-jet and electrographic printing systems are both
liquid based technologies that are driven electronically, they
operate on completely different principles. In this section we
describe the working principles of these printing engines in order
to understand their liquid toner requirements.

2.1   ELECTROGRAPHIC PRINTING

A schematic diagram of an electrographic printing system is shown
in fig. 2. In this printing method a latent electrostatic image is
deposited onto dielectric coated paper (i.e. paper that can hold an
electrostatic charge pattern) by an array of metal stylii which are
selectively discharged according to the electronic input the stylii
receive. The ions that form the latent image are caused by the
dielectric breakdown of the air between the stylii and the paper.
The physics of this process is beyond the scope of this report but
it has been discussed by numerous authors[1-4]. This latent image on
dielectric paper then passes through a development zone in which

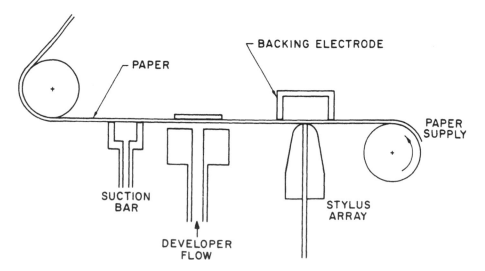

FIG. 2   *Schematic diagram of an electrographic printer.*

liquid developer comes into contact with this electrostatic charge
pattern.

Under the influence of an electric field the particles that
form the liquid developer undergo electrophoresis. Since the
particles are electrostatically charged with the opposite polarity
to that of the latent electrostatic image, the charge on the paper
is neutralised by the developer to give a legible hard copy output.
Once out of the development zone the hydrocarbon fluid on the paper
evaporates rapidly leaving a dry image.

2.2   INK-JET PRINTING

Ink-jet printing is in principle an extremely simple and elegant
method of placing marks on paper. In this printing process a
uniform train of ink droplets is generated by a Rayleigh
instability. These ink droplets are then deflected onto paper to
produce text and graphics.

Numerous variations of ink-jet printers exist although only
two modes of operation appear to have been widely studied. These
are shown schematically in fig. 3. The first method is known as
the drop-on-demand system in which droplets of ink are generated as
needed by a piezoelectric crystal, and are ejected from 20-80 μm
nozzles producing a stream of ink droplets with a velocity of $\sim$ 3 m
sec^{-1}. In order to produce characters on paper the droplets are
assigned the correct trajectory by the writing head. The second
type of jet system is known as synchronous ink-jet printing in
which ink droplets are produced continuously. This is achieved by
pressurising the ultrasonically attenuated jet to $\sim$ 3 x 10^5 Pa which
produces a stream of droplets ($\sim 10^6$ per second) with a velocity of
$\sim$ 20 m sec^{-1}. The ink drops that are to be used to generate
characters are inductively charged and deflected in a high voltage
electric field to a specific position on paper. The uncharged ink
droplets pass undeflected through the electric field to be caught
in a gutter and recirculated through the fluidic circuit.

In the case of electrography, individual electrostatically
charged ink particles (< 2 μm diameter) that are suspended in a

SYNCHRONOUS INK-JET

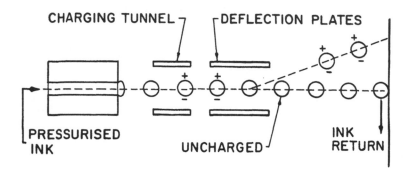

DROP-ON-DEMAND INK-JET

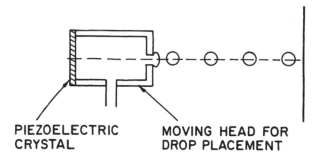

FIG. 3   *Schematic diagram of an ink-jet printing system.*

dielectric fluid are imaged on paper while in ink-jet systems a
macroscopic ink droplet (40 - 150 μm diameter) does the imaging.
The properties of these inks are detailed below.

3.0 DISCUSSION OF THE FUNCTIONAL REQUIREMENTS OF LIQUID DEVELOPERS

One of the major differences between electrographic and ink-jet
inks resides in the fluid medium that each uses. Electrography

must use an insulating liquid such as an aliphatic hydrocarbon, while ink-jet must use a high surface tension fluid such as water. If conductive fluids were used in electrography no development would take place because the fluid would discharge the latent image upon coming into contact with it. On the other hand low surface tension organic fluids cannot be used in ink-jet systems because they do not break-up into a well defined train of droplets. These fundamental physical limitations largely determine the materials that may be used in each technology.

3.1 PROPERTIES OF ELECTROGRAPHIC DEVELOPERS

A schematic diagram of an electrographic toner is given in fig. 4 and indicates that they consist of an insulating fluid, a surfactant to stabilise the particles against flocculation, a surface active charge control agent and a particle. Each of these components must meet specific criteria. These are detailed below.

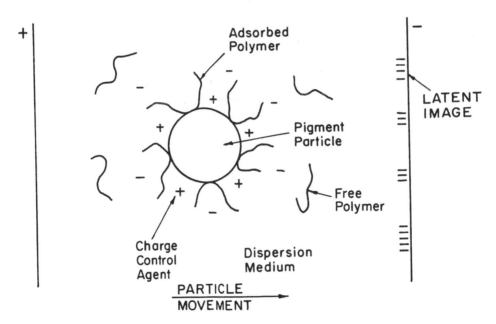

FIG. 4   Schematic diagram that shows the components of a conventional electrographic developer.

Dispersion Medium: This has to be an insulating fluid with a resistivity greater than 10^9 ohm cm so that it does not discharge the latent image. It should also be of low viscosity i.e. less than 2.5 mN.s.m^{-1} to allow for fast migration of the particles through the fluid. The dispersion media that is almost universally used for liquid developers are the Isopar hydrocarbons from Exxon, the G grade being extremely popular for these ink systems.

Surfactant: The purpose of the stabiliser is threefold (i) to help disperse the dry particles in the dispersion medium, (ii) to stabilise the particles against flocculation and (iii) to fix the toner particle to the paper after development of the latent image.

In principle, both high and low molecular weight compounds can be used as the surfactant stabiliser although polymeric materials are usually to be preferred. It could also be ionic or non-ionic in nature, although non-ionic materials are preferred for reasons discussed later in the article. In order for it to be effective it should also have amphipathic characteristics so that part of the molecule can anchor the surfactant to the particle surface. Block and graft copolymers are the surfactants of choice since part of the block copolymer can be made to be soluble in the dispersion medium forming a protective barrier around the particle, while the nominally insoluble block is adsdsorbed onto the pigment surface thereby anchoring the soluble component to the particle surface.

Charge Control Agent: This must impart an electrostatic charge to the particles in order for it to undergo electrophoresis in an electric field. Various postulates as to the mechanism of charging exist, although practically, salts that are capable of undergoing adsorption and ionisation are added to the dispersion medium to impart charge to the particles. The types of materials used as charge control agents are usually of the metal soap variety with numerous examples being quoted in the literature[5-7].

Particle: The function of the particle is as a colourant. The majority of commercially available electrographic inks use pigment particles with carbon black being used almost universally for black liquid developers. The advantages of carbon black are availability, cost and a rich surface chemistry that allows the particles to be readily electrostatically charged in an aliphatic hydrocarbon-based liquid. Organic pigments are used for colored developers. These have historically been more difficult to stabilise and charge in aliphatic hydrocarbon media.

3.2 PROPERTIES OF INK-JET INKS

Fig. 5 illustrates the components that are necessary to formulate a useful ink-jet ink and includes a fluid vehicle and a dye. Other additives such as a biocide, a humectant and a chelating agent are added to prevent operational problems from occurring in the printer. Unlike electrographic inks, which are colloidal dispersions, ink-jet inks are multicomponent solutions[8]. Thus, in principle they could be construed as being simpler to formulate. In practice this has not been found to be the case.

The Solvent: The major requirement of ink-jet inks is that they exhibit a surface tension greater than ~ 35 mN m^{-1}. Since water has a high surface tension, low viscosity, and has the potential of being made conductive, it is traditionally the solvent or cosolvent of choice. Its disadvantages are that it is an ideal medium for

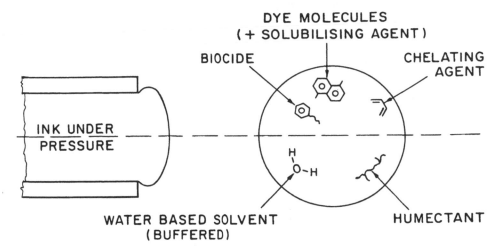

FIG. 5   *Schematic diagram showing the components of an ink-jet ink.*

corrosion of metal parts and for the growth of biological
organisms. In order to prevent biological growth, a biocide is
added to the formulation which is also buffered, usually to an
alkaline pH to minimise corrosion problems. Evaporation of water
is also a problem since it can lead to changes in the dye
concentration within the jet which can cause the dye to
precipitate. This can cause the jet to clog. In order to overcome
this problem a humectant such as an oligomer of ethylene glycol is
added to slow down the evaporation rate.

   Colorant: The colorant used is normally a dye although particles
can be used and this will be discussed in another section. A major
problem with dyes is their limited solubility in water. The dye
concentration for an optically dense image usually needs to be
around 5 wt %. Since it is often difficult to solubilise this
concentration of dye, fairly polar dye-solubilising agents such as
N-methyl pyrrolidone are often added to the solution. Since many
dyes also contain unwanted trace metals these are kept in solution
by the addition of chelating agents.

   With the addition of so many additives, what was conceptually
a simple ink turns out to be fairly complex. The interactions that
occur in solution are not well understood. Thus ink-jet
formulation technology remains very much an intuitive art.

4.0 PHYSICOCHEMICAL AND IMAGING CHARACTERISTICS OF LIQUID DEVELOPERS

The physico-chemical characteristics required of ink-jet and
electrographic developers are shown in Table 1. Each of these
technologies is discussed separately.

4.1 ELECTROGRAPHIC TONERS

In order to be functionally useful the particles must be
colloidally stable, electrostatically charged and image in a

reprographic device. Surfactants play a major role in this technology since they are used to colloidally stabilise the particles and also to charge them.

4.1.1 Colloid stability The colloidal particles constituting the developer should remain as discrete entities over the lifetime of the toner. This is not an insignificant problem since by definition colloidal dispersions are almost always thermodynamically metastable. If naked colloidal particles are dispersed in a fluid medium, they coagulate quickly due to the attractive van der Waals forces (V_A) that exist between the particles. In order to stabilize the dispersion, a repulsive potential (V_R) must be introduced between the particles such that $V_R > V_A$. This can be achieved either by introducing a steric barrier, or by introducing a repulsive electrostatic potential between the particles. The total particle interaction, V, of two particles in a dispersion is to a first approximation, given by[9]

$$V = V_A + V_R \tag{1}$$

and

$$V_R = V_E + V_S \tag{2}$$

where V_E is the double layer interaction contributed by the charge control agent and V_S is the contribution of the steric stabiliser to the particle interaction. In theory, the charge control agent can generate electrokinetic potentials greater than 50 mV, which translates into a repulsive interaction, V_E, that is larger than the attractive van der Waals interaction, V_A. Therefore, theoretically there is no need for a steric stabilizer. *However, since we prefer to use the electrostatic charge to facilitate transport through the fluid medium, we choose not to rely on this technique to colloidally stabilise the particles but to utilise a steric stabiliser for this purpose.* The value of the electrostatic charge imparted is also dependent upon the application for which it

is to be used and can thus, to a particle vary dramatically. This
is another reason for separating the colloidal stability and
charging functions in an electrographic developer.

A first order approximation for the V_A between two particles
is given by the expression[9]

$$V_A \sim \quad - \frac{A^{*}a}{12H} \qquad (a \gg H) \qquad (3)$$

where A^{*} is the effective Hamaker constant which takes account of
the medium in which the particles are dispersed. The radius of the
particles is denoted by a where H is the surface-to-surface
separation. Eq. 3 indicates that V_A is directly proportional to a.
This means that the larger the particles the more difficult they
are to stabilise. This is illustrated for carbon black particles
in Fig. 6a.

The adsorption of a polymeric surfactant from solution onto
the surface of a pigment gives rise to a repulsive steric barrier
between the particles. It has been found that steric stabilisation
forces are of a shorter range than attractive forces[10] and increase
rapidly as soon as the steric barriers come into contact. Numerous
workers have contributed to the theory of steric stabilization with
the various expressions being able to be written in the simple
form[10]

$$V_S = B(1/2 - \chi)aSkT \qquad (4)$$

where B is a function of the molecular parameters of the steric
stabiliser and the dispersion medium and is always a positive
quantity; S is a function that accounts for the distance dependence

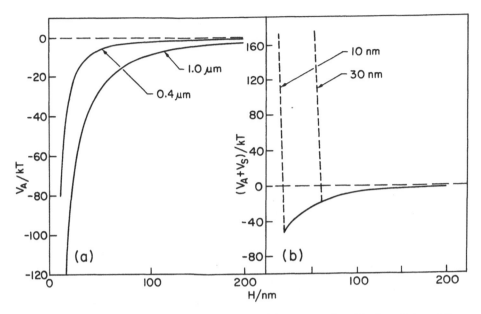

FIG. 6   *(a) Shows the attractive potential (V_A) between carbon black particles while (b) shows the effect of steric interactions (V_s) on the interaction between carbon black particles.*

of the particle interaction and χ is a parameter which is a measure of the antipathy between the stabilising polymer and the dispersion medium. Equation 4 predicts that the dispersion will be stable when Vs is positive, which means that $\chi < 1/2$. When $\chi > 1/2$, Vs is negative and the system flocculates. The condition $\chi = 1/2$ corresponds to the theta (θ) point of a polymer solution. Therefore, the θ point for the sterically stabilising polymer in the dispersion media represents the theoretical limit of stability of the nonaqueous dispersion[10]. This analysis is based on the assumption that the polymeric stabiliser *does not* desorb from the particle surface. For pigment based electrographic inks, where we rely on effective adsorption of the stabiliser, desorption is always a potential problem of which we should be aware. Some of the polymers that are used as stabilisers for electrographic inks are shown in fig. 7. While these materials are useful they are not

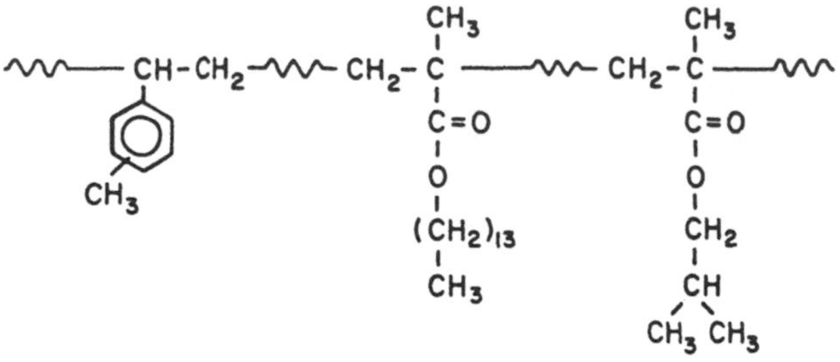

$$CH_3 \quad\quad CH_3$$

```
∿∿—CH-CH₂—∿∿—CH₂-C——∿∿—CH₂-C——∿∿
    |                    |              |
    ◯                    C=O            C=O
    |                    |              |
   CH₃                   O              O
                         |              |
                      (CH₂)₁₃          CH₂
                         |              |
                        CH₃            CH
                                     CH₃ CH₃
```

PLIOLITE OMS (GOODYEAR)

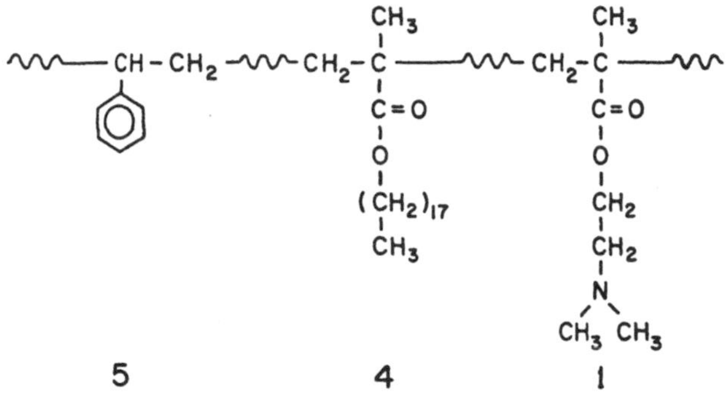

$$CH_3 \quad\quad CH_3$$

```
∿∿—CH-CH₂—∿∿—CH₂-C——∿∿—CH₂-C——∿∿
    |                    |              |
    ◯                    C=O            C=O
                         |              |
                         O              O
                         |              |
                      (CH₂)₁₇          CH₂
                         |              |
                        CH₃            CH₂
                                        |
                                        N
                                      CH₃ CH₃
     5             4                    1
```

ALKANOL DOA (DuPONT)

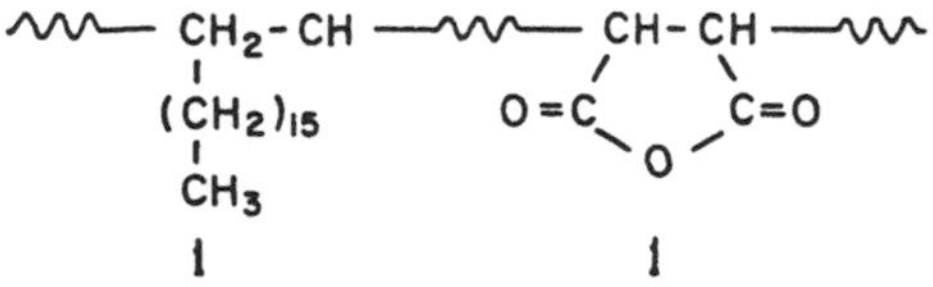

```
∿∿—CH₂-CH——∿∿——CH-CH——∿∿
     |              /  \
   (CH₂)₁₅     O=C      C=O
     |             \  /
    CH₃             O
     |              |
     1              1
```

PAI8 (GULF)

FIG. 7   *Shows some commercially available polymers that have been used as stabilisers in electrographic printing.*

ideal. A discussion as to the ideal surfactant will be given in section 5.

As was mentioned earlier, the range over which steric interactions occur are extremely short. For the steric barrier to be effective it should effectively screen the attractive forces between the particles. A good stabiliser needs to be of sufficient thickness, i.e., of a sufficient molecular weight, to stop the particles flocculating in a potential energy minimum which has an energy well greater than a few kT. This is an important practical consideration in designing functionally useful electrographic toners and is illustrated in fig. 6b for different barrier thicknesses. It should also be borne in mind that the electrostatic charge on the particles will enhance the particle stability although we do not rely on this mechanism to stabilise the particles. This contribution can be thought of as a bonus but should not be relied upon in formulating electrographic inks.

In direct electrographic printing the concentration of particles in the developer is usually of the order of 1% by weight. In order to reduce gravitational settling of electrographic inks the particles are usually attrited until they are submicron in size. The problem of their settling behaviour has been discussed in more detail elsewhere[5].

4.1.2 *Electrical characteristics* In order for the toner to image it must be transported under the influence of an electric field. Consequently, the particle must be charged. It is usual in electrographic toner technology to measure the charge/mass (Q/M) ratio of the ink and to relate this with imaging performance in a printer[5]. Measurements of this ratio have been described before[5] and can be related to fundamental electrokinetic parameters of the dispersion[5]. Thus

$$\frac{Q}{M} = \frac{3\varepsilon\zeta}{a^2\rho_p} \qquad (5)$$

where ε is the dielectric constant of the dispersion medium, ζ is the zeta potential and ρ_p is the density of the particles. Equation (6) can alternatively be written as [5]

$$\frac{Q}{M} = \frac{9\eta u}{2a^2\rho_p} \qquad (6)$$

where u is the electrophoretic mobility of the particles and η is the viscosity of the newtonian dispersion medium. Thus, Q/M is proportional to u/a^2, which is a function of the electrophoretic mobility of the system. Qualitatively, we can say that for rapid development of the image, Q/M should be large. However, the optical density of the developed image depends upon the number of particles deposited per unit area, and this will obviously be dependent upon the Q/M value. Therefore, there is a trade-off between development speed and optical density. This should be borne in mind in designing a commercial toner. In commercial printing devices the Q/M ratio can vary from 10 - 1000 μC g^{-1} depending upon the application for which it is to be used. For instance, where images of a high optical contrast are required such as in medical imaging systems the Q/M value would be ~ 100 μC g^{-1} while in high speed printing applications the Q/M value would more likely be ~ 800 μC g^{-1}.

Of perhaps more fundamental importance for this ink technology is understanding how to effect and control the electrostatic charge on a particle in a dielectric medium using a surfactant. This is a subject which is not well understood at the present time. It is thought that particles in dielectric media can be charged in numerous ways, the most useful being:

1)   by acid-base interactions
2)   by adsorption of ionic surfactants.

A subject of considerable interest in recent years is the effect of acid-base interactions. Fowkes has published extensively on this subject[11] and only a summary of his ideas will be presented here. The mechanism of acid-base interactions giving rise to a charged particle is shown schematically in fig. 8a. The first step involves the adsorption of a surfactant (which is depicted as having a basic character) onto the surface of a particle (which is shown as having an acidic surface chemistry) together with transfer of a proton from the acidic sites on the particle to the basic

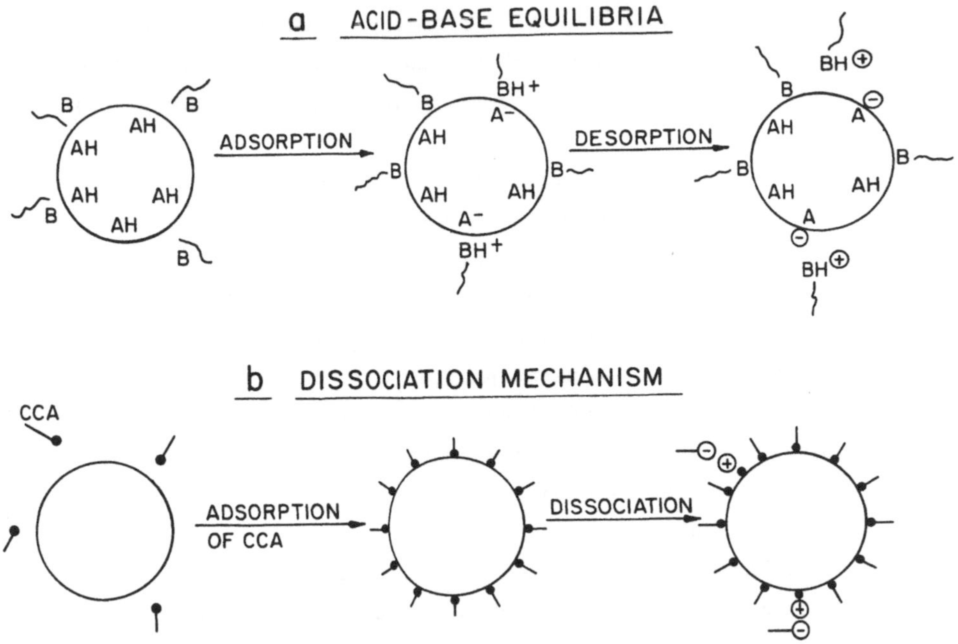

a   ACID-BASE EQUILIBRIA

b   DISSOCIATION MECHANISM

FIG. 8   *Shows schematically two proposed mechanisms of generating electrostatic charge on particles in dielectric media.*

group of the adsorbed surfactant. The second step required for particle charging is desorption of the proton-carrying polymer from the surface into the solution, leaving a negative charge on the particle. Evidence for this mechanism was provided using ^{14}C tagged surfactant materials to follow the adsorption-desorption process. While this mechanism certainly has an intuitive appeal, it is by no means certain that it is correct or is the only mechanism. A more conventional view is shown in Fig. 8b in which a surfactant is adsorbed at the solid-liquid interface, followed by dissociation of an extremely small fraction of the adsorbed molecules to leave a charged particle. It has been estimated[5,12] that about one molecule in ~ 10^4 adsorbed molecules dissociates in this process. Considering the fundamental importance of understanding the mechanism of the electrostatic charging of particles, it is perhaps surprising that so little effort has been directed towards this subject.

Most of the work emanating from this laboratory[5-7] has
involved metal soaps as the charge control agent, since we have
been interested in imparting a positive charge to particles, rather
than a negative charge. These materials have been found to give
rise to relatively large equivalent conductances. It has not been
established whether the conduction is due to simple ions or to some
multiple ionic species that occur in solution. In the systems we
have studied it is always the cation that is in the adsorbed state
while the anion is in solution in equilibrium with the non-adsorbed
species in solution. A major problem with these materials is that
they are not well characterised from a structural chemistry
viewpoint and we have little knowledge regarding their purity.
This makes an understanding of the mechanism of electrostatic
charging difficult to achieve.

From measurements of Q/M it is possible to obtain some rather
crude estimates of the electrokinetic parameters of the particles[5]
using eq. 6 and 7. These measurements have shown that zeta
potentials range from 50 - 250 mV, while electrophoretic mobility
values are in the range of $10^{-8} - 10^{-9} m^2 V^{-1} s$. These zeta
potential values are extremely large and indicate that particles in
dielectric media can in principle be charge-stabilised. However,
as we have mentioned previously, we prefer to sterically stabilise
particles while allowing the charge to facilitate transport of the
particles in the development zone.

4.1.3 Fixing and imaging properties of electrographic developers   The adhesion
of the electrographic image to paper must be adequate to avoid rub-
off of the image on handling. Three factors are thought to
contribute to the fixing characteristics (a) the roughness of the
paper surface (b) the properties of the dielectric layer and (c)
the film-forming and wetting characteristics of the polymeric
stabiliser/fixant on dielectric paper. It has been found that
electrographic inks were better fixed to dielectric paper with
rougher surfaces than those with smoother surfaces. A tentative
explanation is that on rougher surfaces the toner can become
embedded within the paper structure thereby making it less

accessible upon handling. However, no systematic study of the
fixing properties of electrographic toners has been reported in the
literature and much work remains to be done in this area.

The pigment based inks we have prepared have all been found[5,6]
to have adequate fixing characteristics but could be smeared when
continuously rubbed. This is obviously a major disadvantage in an
office printer since transfer of the ink from paper to hands or
clothing is unacceptable in a modern office environment.

The optical density or absorbance of the toned black image
should be in the range 1.1 - 1.4 and this can readily be obtained
with carbon black-based developers. The Q/M ratio of the toners
has to be carefully controlled, since it has been found that if
excess ions are in the developer they compete with particles to
discharge the latent image in the development zone[13].

The resolution in electrographic images can be exceptionally
good and 200-line pairs per millimeter (1p/mm) has been achieved.
This high resolution toning process makes electrographic an
attractive proposition for technologies where resolution is
important, as in micrographics or color printing.

4.2   INK-JET DEVELOPERS

In comparison to electrographic developers, ink-jet inks appear to
be extraordinarily simple since the only physicochemical properties
of interest are the surface tension (or more properly the dynamic
surface tension), the viscosity, the conductivity and the pH of
these complex multicomponent solutions.

4.2.1 *The surface tension*   In ink-jet printing, ink droplets are ejected
from one or more nozzles onto paper forming alphanumeric
characters. Such imaging systems employ principles of physics
which have been recognized for the last century. It is well known
that a liquid issuing from an orifice attempts to break up at a
resonant frequency producing droplets. Since the jet is prone to
vibration and noise, the liquid breaks up in a fairly random
fashion. If an external perturbation is applied to the jet close
to its resonant frequency, the noise and vibration are suppressed

and the jet produces extremely uniform droplets. The problem of the stability of an infinitely long circular cylinder of incompressible fluid was first analysed by Rayleigh[14].   The jet velocity, separation length of the droplets, drop size and stream stability are greatly affected by the surface tension and the viscosity of the ink. It has been shown that if the surface tension of the fluid is greater than about 35 mN m^{-1}, then a well defined stream of droplets is formed. As the surface tension becomes smaller than ~ 35 mN m^{-1} it is found that it is more difficult to form the well defined droplet size that is demanded by this technology. Consequently, a minimum value for the surface tension of ~ 35 mN m^{-1} is usually specified. Because water has a high surface tension it is often used as the base fluid for ink-jet inks. The choice of dyes, humectants and other additives is largely governed by their surface behaviour. If they are highly surface active, they will be unacceptable. One of the areas of concern in conventional ink-jet inks is the limited solubility of the dyes that are used. Among the various ways of increasing this mutual solubility is by the addition of another material which will increase the solubility of the dye. These additives may operate by a number of different mechanisms which include (a) cosolvency (b) formation of a pseudophase such as a micelle or surfactant phase and (c) formation of soluble complexes. The minimum surface tension requirement for ink-jet inks limits the choice of solubilising materials for this technology.

4.2.2 Rheological properties of inks Rayleigh originally attributed the breakup of a flowing liquid into droplets entirely to dynamic surface tension effects[14]. The analysis for viscous liquids and for elastic fluids was subsequently carried out by Weber[15] and Goldin et al[16] respectively. These analyses showed that the growth rate of the disturbances causing drop formation are very viscosity dependent. Therefore, the drop-forming characteristics of the ink are sensitive to the viscosity. Unfortunately, the shear rates of ink-jet liquids are of the order of 10^6 sec^{-1} and little is known

of the rheological behaviour of solutions in this regime, although
it is important in other practical applications such as lubrication
and drag reduction. It should also be mentioned that as some ink-
jet inks contain polymers, they will likely impart elastic
properties to the fluid. These would be pronounced at high shear
rates and could possibly have a dramatic effect on the
characteristics of the drops that are formed and consequently on
copy quality. For a useful ink-jet ink the viscosity of the liquid
should be in the 1-10 mN.s.m^{-2} range and should preferably be
Newtonian in character. Many commercial ink-jet inks are non-
Newtonian and this can be attributed to the "structure" within the
fluid caused by the interactions in these complex mixtures.

4.2.3 Conductivity and pH of inks The conductivity of the ink is only
important if it is to be used in a synchronous ink-jet printer.
Here, the ink droplets are deflected to the required position on a
piece of paper. In order for this to occur the ink droplet has to
be charged. This is carried out by inducing a charge on the
droplets which means that the fluid has to have a conductivity
greater than ~ 10^{-3} (ohm cm)$^{-1}$. The conductivity of these inks is
not usually a concern because the types of additives that are used
in these developers such as biocides and chelating agents are often
ionic in character and are capable of making the ink conductive.
On the other hand, conductivity is of no consequence in drop-on-
demand type inks because the droplets do not have to be charged in
order to be steered to the correct place on the paper.

   Adjusting the pH of the ink to the correct value is often a
more important consideration. If the ink comes into contact with
stainless steel parts the pH is normally adjusted to the range 9 to
11 since this corresponds to the region of minimum corrosivity for
stainless steel in aqueous fluids[8]. To prevent a drop in the pH
with time due to the absorption of carbon dioxide, the ink is
usually buffered with sodium carbonate[8]. If the printer is made
entirely from plastic rather than metal, then the pH requirement is
less severe and offers the ink formulator the opportunity to pursue
a wider range of materials.

4.2.4 Imaging characteristics of ink-jet printers The resolution obtainable from ink-jet printing is very much dependent upon the size of the droplet ejected from the nozzle and the quality of the paper that is used. The resolution obtainable on present products is ~ 8 lp/mm. However, if one could put down 20 μm spots on paper a resolution of ~ 50 lp/mm is theoretically possible. A major cause of image degradation in ink-jet systems is the image feathering that takes place along the paper fibres of the paper[17]. This is illustrated in Fig. 9. Ideally, the spot formed should be well defined. In reality an irregularly shaped spot is formed with a consequent loss in image resolution. The degree of spreading depends upon the paper used. Consequently, one of the challenges in ink-jet technology is to formulate an ink that will spread as little as possible on as wide a range of papers as possible.

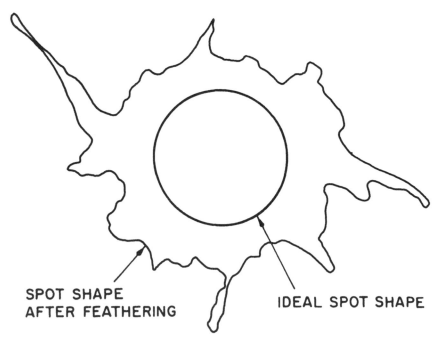

SPOT SHAPE
AFTER FEATHERING   IDEAL SPOT SHAPE

FIG. 9   *Shows schematically the effect of ink feathering on the ideal spot size leading to image degradation.*

The other imaging criteria of importance are the image density which for black should be greater than ~ 1.2, and the drying time which should be of the order of seconds. While dense images are readily obtainable, short drying times are more difficult, especially for solid areas, since evaporation of water is slow and requires considerable energy. Thus in ink-jet technology an absorptive paper is an important requirement. Finally, the waterfastness of the image is a major issue[8]. Since water soluble dyes are used it is difficult to achieve complete waterfastness, although major advances have been achieved with the use of polymer additives that anchor the dye to the paper.

5.0 FUTURE DIRECTIONS IN LIQUID DEVELOPERS

In the previous sections we have outlined the materials used and the physicochemical properties of liquid developers for use in electrographic and in ink-jet printing. At present, the design criteria and the materials used for these technologies are radically different. It is likely that the design rationale behind these liquid developer systems will converge and this theme will form the basis of this discussion.

5.1 SURFACTANTS IN ELECTROGRAPHIC PRINTING

Surfactants have a dual function in electrographic inks: (a) they stabilise the colloidal dispersion of particles and (b) they impart charge to the particles.

In order to stabilise particles it is widely accepted that well-tailored block and graft copolymers are necessary, where one part of the copolymer forms the "anchor" to the particle surface and the second moiety provides the "solvated steric barrier". The types of molecular architectures that would be useful to stabilise particles in media of low dielectric constant are shown schematically in Fig. 10. Recently, Hampton and MacMillan[18] and Jakubauskas[19] have discussed the use of block and graft copolymers

BLOCK COPOLYMER STABILISERS

GRAFT "COMB" STABILISERS

FIG. 10   *Shows schematically block and graft copolymers of different molecular architectures that would be useful as steric stabilisers.*

for non-aqueous applications. In particular, they point out the difficulty of stabilising the non-polar surfaces often found in organic pigments. Hampton et al[18] discuss the use of surface-modifying agents, which can provide good anchoring to non-polar surfaces. These surface modifying agents are pigment derivatives containing polar groups. They are adsorbed on the surface of the pigment by virtue of their physicochemical similarity to the pigment and provide anchoring sites for the block or graft copolymers. While this is an interesting idea, it would appear to be surface specific. More importantly for electrographic printing, it could conceivably hinder the color and charging characteristics of such liquid toners.

Surfactants are also used as charge control agents in liquid developers. At present a charge control agent for a specific developer is found using an iterative process. This occurs because we do not fully understand the mechanism of producing an electrostatic charge at a pigment surface. Until a greater understanding of charging is achieved, it is difficult to describe the types of surfactants that will be useful for this function.

Ideally, the role of the dispersant and of the charge control agent should be embodied in one molecule and this would seem to be an important strategic long term objective.

5.2   USE OF SURFACTANTS IN INK-JET INKS

The conventional ink-jet inks that are presently used are complex multicomponent solutions[8]. They contain dye(s), polymers, solubilising agents, chelating agents and biocides. These complex interactions are incompletely understood and can and do lead to failures within the printer. It seems clear that a new approach to formulating such materials is required that allows for a high surface tension, higher image resolution and waterfastness. The question that needs to be asked is "how can a waterfast oil soluble dye be molecularly dispersed in a water-based ink?" It seems that this could be accomplished by molecularly dispersing a dye within a colloidal structure such as a latex particle, an emulsion or microemulsion or within a surfactant aggregate. We will discuss the use of latex particles in another section. To date, emulsions[20,21] have been patented for ink-jet inks but we are not aware of any published work on microemulsions or surfactant aggregates. An obvious problem with microemulsions is that alcohols are often used as cosurfactants. It would therefore appear to be difficult to formulate a microemulsion based ink with a surface tension that is large enough for the ink to break up into ink droplets. The use of surfactant aggregates would seem to be a more useful avenue to explore. In this approach one would want to molecularly incorporate a dye within the aggregate structure. This would probably require specially designed materials to control the dye concentration and colloidal size while still observing the physicochemical characteristics listed in Table 1.

5.3   LATEX PARTICLES AS LIQUID DEVELOPERS

Latex technology is now fairly well developed and techniques are available of making polymer particles in a variety of liquid media with a controlled particle size and size distribution. If these

TABLE 1 *Physicochemical Requirements of Liquid Developers*

Property	Electro-graphic Developer	Ink-Jet Ink	Comment
Surface Tension/mN.m⁻¹	--	> 35	*Ink-jet need a stable jitter free stream of droplets*
Conductivity/(ohm/cm)⁻¹	$< 10^{-9}$	$> 10^{-3}$	*Drop on demand printer has no conductivity requirement*
Viscosity/mN.s.m⁻²	$\leqq 2$	*1 - 10*	
Shelf Life Stability/months	*18*	*18*	*Materials should not age and/or change with time*
Drying time/sec	*< 10*	*< 10*	*Drying of image is more difficult for ink-jet than for electrography*

particles can then be colored they could form the basis of either electrographic or ink-jet developers. A method of making particles that we have explored in some detail is known as dispersion polymerisation[22]. A schematic of the dispersion polymerisation process is shown in Fig. 11. In this polymerisation process a monomer, which is soluble in the dispersion medium is polymerised in the presence of an amphipathic block or graft dispersant. The other requirement of this process is that the polymer that is formed be insoluble in the dispersion medium. Consequently, after polymerisation to a specific chain length the polymer will precipitate from solution. These precipitated polymer chains form the nuclei of latex particles which are then stabilised by the amphipathic copolymer, allowing the nuclei to grow into discrete sterically stabilised polymer particles. The advantage of this technique is that particles can be polymerised in both aliphatic hydrocarbon media, such as Isopar G or in water based mixtures. A micrograph of some dispersion polymerised polymer particles is

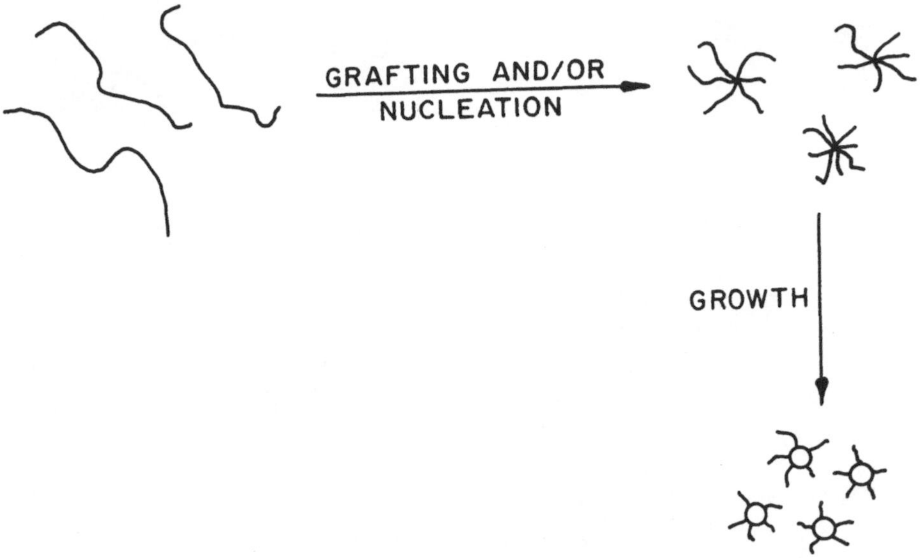

GRAFTING AND/OR
NUCLEATION

GROWTH

FIG. 11   *Shows schematically the dispersion polymerisation process.*

shown in Fig. 12. The advantage of these sterically stabilised
particles is that they are thermodynamically stable[10] since the
stabilising chains are terminally attached to the particle surface.
Once these particles have been colored they can form the basis of
either ink-jet or electrographic inks.

For electrographic toners the colored particle has to be made
in an aliphatic hydrocarbon medium, dyed and electrostatically
charged. It has been found[7,23] that numerous metal soaps are
useful in charging latex particles. A major advantage of this
approach to making developer materials is the control that can be
exerted over particle size, the colloidal stability of the
materials and their film forming characteristics. This technique
shows great promise for being able to make electrographic inks with
specified, well controlled properties, but more research is
necessary before they become a commercial reality.

A major thrust in ink-jet technology is to make the inks less
paper dependent. A method of achieving this end is to make

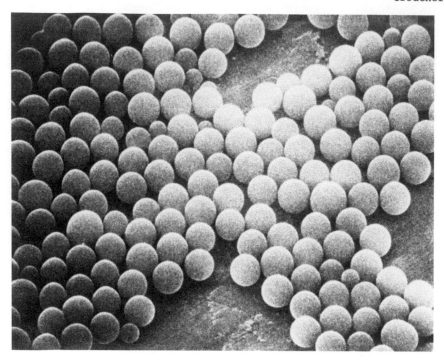

FIG. 12   *This shows a scanning electron micrograph of dispersion polymerised latex particles that could form the basis of either electrographic or ink-jet inks.*

particulate ink-jet inks[24,25]. Dispersion polymerisation in a water based fluid is one method of achieving this objective[25]. Control over particle size is a primary objective with very small particles (a few hundred angstroms) being desired. This will mean that the particles will not settle and clog the printing head. A further advantage of sterically stabilised particles, unlike electrostatically stabilised emulsion polymerised particles, is that they are stable[10] over a wide pH range, which is a useful property to build into these inks. This concept for ink-jet inks has now been demonstrated[25], but further development work is required. One of the major requirements for a successful dispersion polymerisation is the commercial availability of steric stabilisers. Such block and/or graft copolymers with the required properties are not at the present time readily available. This will probably retard the development of latex based ink-jet and electrographic developers.

REFERENCES

1. J. A. Dalquist and I. Brodie, *J. Appl. Phys. 40*: 3020: (1960).

2. V. Novotny in "Colloids and Surfaces in Reprographic Technology", Ed. M. L. Hair and M. D. Croucher, *ACS Symposium Series 200*: 281 (1982).

3. I. C. Roselman and W. Tait, *SPSE Conference, Washington, D.C.* (1977).

4. R. B. Crofoot and Y. C. Cheng, *J. Appl. Phys., 50*: 6583 (1979).

5. M. D. Croucher, S. Drappel, J. Duff, K. Lok and R. Wong, *Colloids and Surfaces, 11: 303* (1984).

6. M. Croucher, S. Drappel, J. Duff, G. Hamer, K. Lok and R. Wong, *Photogr. Sci. Eng., 28*: 119 (1984).

7. M. D. Croucher, K. P. Lok, R. W. Wong, S. Drappel, J. M. Duff, A. Pundsack and M. L. Hair, *J. App. Poly. Sci., 29*: 593 (1985).

8. C. T. Ashley, K. E. Edds and D. L. Elbert, *IBN J. Res. Develop., 21*: 69 (1977).

9. E. J. Verwey and J. Th. G. Overbeek, in "Theory of the Stability of Lyophobic Colloids", *Elsevier, Amsterdam* (1948).

10. D. H. Napper, "Polymeric Stabiliation of Colloidal Dispersions", *Academic Press* (1983).

11. F. M. Fowkes, F. W. Anderson, R. J. Moore, H. Jinnai and M. A. Mostafa, in "Colloids and Surfaces in Reprographic Technologies", Ed. M. L. Hair and M. D. croucher, *ACS Symposium Series, 200*: 307 (1982).

12. V. Novotny, *Colloids and Surfaces, 2*: 373 (1981).

13. J. M. Duff, R. W. Wong and M. D. Croucher, in "Surface and Colloid Science in Computer Technology", in press.

14. Lord Rayleigh, *Proc. Lond. Math. Soc., 10*: 4 (1879); *Proc. Roy. Soc., 29*: 71 (1879).

15. C. Z. Weber, Agnew, *Math. Mech., 11*: 136 (1931).

16. M. Goldin, J. Yerushalmi, R. Pfeffer and R. Sinnar, *J. Fluid. Mech., 38*: 689 (1969).

17. J. Oliver, in "Surface and Colloid Science in Computer Technology", in press.

18.   J. S. Hampton and J. F. MacMillan, *American Ink Maker, 63*: 16 (1985).

19.   H. L. Jakubauskas, *J. Coatings Technology, 58*: 71 (1986).

20.   Japanese Patent J56166274-E05 to Ricoh (1980).

21.   European Patent EP-36790-D41 to Hotchkiss-Brandt (1981).

22.   M. Croucher, Preprint Wingspread Polymer Colloids Workshop, to be published (1986).

23.   M. D. Croucher, J. M. Duff, M. L. Hair, K. P. Lok and R. W. Wong, *U. S. Patent 4,476,210* (1984).

24.   Y. Yao, *U. S. Patent 4,246,154* (1978).

25.   C. Ober, M. L. Hair, R. W. Wong and M. D. Croucher, to be published.

Electronic Printing
Group Discussion I

C. Herve, Syracuse University, rapporteur

Even though ink jet printing is in its teens, it still requires novel concepts. The idea was originally demonstrated and
put into practice by IBM in the mid 1970s. Since then everyone
has followed their method, using dye-containing, water-based
fluids. Unfortunately, this method is very paper dependent, and
only a very limited range of paper types can be used. All of the
reprographic companies are trying to resolve this problem, by
obtaining water fastness and stopping image degradation, in order
to create an ink which will image on a broad range of papers.

Because of the tremendous limitations of the water-soluble
dye approach, the leading theme throughout the workshop was the
idea of starting from an oil-soluble dye and encapsulating it in a
colloidal body of submicron size in order to avoid clogging the
20 μm orifice. This water-based stabilization could possibly be
made out of small latex particles or out of a less obvious colloidal body, such as a surfactant aggregate. The use of ionomers
was also mentioned; they would give water fastness and solubility,
but then research would have to be done in order to observe what
would happen to the various species and solutions and how they
would interact. An alternative to this "tough way to go" would be

to start from a more rational approach: to design a particle
which would not migrate on the paper and then to incorporate a
colorant into it. After this is done, stability and non-settling
behavior would have to be obtained. Once this is accomplished,
other additives could be used to reach the right conductivity.

Image degradation stems from lateral migration on the paper.
By solving this problem, the image could be maintained regardless
of the structure of the paper. Paper possesses specific functional
groups on the surface which enable the "dye-containing particle"
to remain adsorbed on the paper. In order to enhance these interactions, particles could also be formed with functional groups
attached to their surfaces which would provide a greater attrac-
tion, and thus diminish the risk of migration.

The pH conditions for the stabilization depend on whether or
not it comes into contact with metals, the least corrosive conditions corresponding to a pH between 9 and 11. Even though electrostatic stabilizations exist at these pHs (using low polarizing ions
as counterions in order to build macroions out of the particles),
steric stabilizations are usually preferred. In addition, there is
also a very good U.S. patent filed by Fuji Photofilm which is
basically blocking all patents on electrostatically stabilized
latex technology for ink-jet inks.

Numerous patents, mostly Japanese, have also been filed on
pigment-based ink-jet inks. However, it is still very difficult
to obtain anything which is small enough (< 1000 angstroms) and
yet still non-settling. Reliability is a major issue as the
orifice must be open at all times and, therefore, it cannot become
clogged with ink jet materials.

On the other hand, considering the tremendous shear experience
that some of the materials undergo, it would be interesting to use
a substance which would be monomeric in the stabilization, but
would start to polymerize when passing through the orifice, due to
shear thickening. This solution would avoid issues such as
flocculation in the reservoir and shear stability.

The necessity of having very well-controlled droplet size restricts the surface tension requirements, which in turn limits the use of surfactant aggregates, such as micelles and microemulsions. In addition, the long term stability limits the use of vesicles, even when polymerized.

A great deal of work has been done on magnetic stabilizations during the past 30 years, and IBM owns several patents on magnetic ink jet printing using magnetite (Fe_3O_4) particles. This technique is a variation of continuous technology using a magnetic field. Even though it is very color limited, the technique offers better image control since the signal of the particle does not vary, as opposed to the more difficult task of keeping the charge to mass equivalent necessary in other techniques.

The future of ink jet inks resides in the production of smaller, monodispersed, colloidal entities (200 angstroms in diameter) which remain stable over long periods of time (18 months), with attention also being given to the strict requirements of viscosity, surface tension, and age degradation. No matter which starting entity is used - latex or very small pigments, solid particles or surfactant aggregates - each has its own advantages and drawbacks. In addition, laser printing has already superceded ink jet printing in most areas. The only area where ink jet technology still plays, and will continue to play, a major role is in color printing. At the moment, even though color printing occupies a major place in today's world, it is not a major reprographic tool. Hard copies of increasingly used color displays are needed, and ink jet is one of the few technologies which is able to provide them. However, ink jet is very paper-limited, and it will never become a major reprographic technology unless it gains more latitude and performs better than it does at the present time.

Electronic Printing
Group Discussion II

M. N. Memering, Pfizer Pigments, Inc., rapporteur

Overview: This session focused primarily on the role of surfactants in two specific liquid developer systems: electrography and ink jet printing. These two systems are similar in that they consist of a formulated suspension of pigment particles, surfactants, and solvent. The primary distinction lies in mode of transport to the paper as well as in the mechanism of drying. In each case, the primary role of the surfactant is to yield a stabilized suspension relative to sedimentation. The electrographic developer is based on an isopar hydrocarbon solvent whereas ink jet inks are currently multi-component solutions in water; the mechanism for drying therefore involves solvent evaporation and substrate absorption of water, respectively. The electrostatic transport mechanism involved in electrography necessitates incorporation of a second surfactant to control interfacial charge of the dielectric media. The latter phenomenon currently lacks fundamental understanding. Since knowledge of mechanisms of charging processes at interfaces is vitally important, not only for electrography, but for dielectric fluid applications in general, basic and applied research of such phenomena should be intensified in the future.

Comments on other non-impact printing systems were inter-
spersed throughout the discussion, including such dry processes
as classical xerography, laser printing, magnetography and
standard photography in addition to liquid magnetic developers.
Though attention was primarily focussed on the intrinsic technical
problems peculiar to each reprographic system, several extraneous
factors were noted as direct influences on future technical direction of reprographic development. It is ironic that with respect
to fundamental understanding, liquid developers are presently in
worse shape than dry toner systems despite their being simpler in
theory. Several untried ideas were proposed and discussed. Often
the primary obstacle to demonstrating such ideas is psychological
rather than scientific in nature.

Electrography: Electrographic developers currently consist of a
formulated mixture of carbon black or organic pigment, isopar
hydrocarbon solvent and two surfactants: one for colloid stability
and one for charge control. Since the system consists of random
copolymers, finding surfactants which consistently assure shelf-
life stability is often a problem. Dr. Croucher proposed the concept of using block and graph copolymers in which one part of the
copolymer would be anchored to the particle and the second part
would be soluble in the isopar. The number of soluble surfactants
is currently limited and certain of these materials also lack the
performance characteristics needed for the application, such as
adhesion to the paper itself.

A second requirement of the surfactant is control of charge
for the electrostatic transport process. Currently there is no
real understanding of the mechanism of imparting charge on particles so that the current knowledge of charge derivation in
dielectric fluids is empirical in nature, requiring iterative
processes to yield desired results for specific applications. For
the electrographic process, not only the charge on the particle,
but also the background conductivity must be under control.
Dr. Croucher referred to one of the visuals presented in the morning wherein a plot of optical density versus conductivity goes

through a maximum. Thus, the operating range involves a relatively narrow window which varies from surfactant to surfactant.
Currently, identification of optimal surfactants involves, not
art, but black magic.

The phenomena of ion mobility in dielectric fluids has spawn-
ed a relatively large industry ($20MM/yr. in U.S. for liquid
developers alone). Although electrography is well established,
better understanding of charge transport phenomena would beget
better control so that electrography could be expanded to even
greater number of applications than the presently, rather specialized, ones. Dr. Croucher suggested that surfactant functions be
decoupled; viz., separate transport from colloid stability. Since
liquid toners have different electrical characteristics, reliance
on electrostatic stability limits flexibility with respect to
applications. Ideally one wants to be able to vary charge on
different particles, such as can be effected in aqueous media by
pH adjustment. For example, in mammography very low charges
are required, so low in fact that achievement of colloid stability
is difficult; whereas for ordinary printing devices charge is very
high and stability is easily achieved. Dr. Croucher proposed
therefore a terminal attack in which a base formulation is produced, shipped to different departments within the company for
dye incorporation, and then a charge control agent added at the
very end. Thus commonality is sought in the early stages and only
in the final step would the liquid developer be differentiated in
accordance with its application. This idealized scenario requires
the addition of a material such as a surfactant that imparts a
different charge depending upon how much is added; unfortunately,
Dr. Croucher had to admit that he had no clue as to how to do this.

The need for fundamental understanding of mechanisms for imparting charges at interfaces in dielectric fluids is not peculiar
to electrographic liquid developers. Indeed the need for such
understanding is quite obvious in such systems as motor oils,
drilling fluids, hydraulic pumps, etc. In such areas, controlled
conductivity is a major requirement because if electrostatic

charge is allowed to build up, explosive situations arise. Additives to such fluids (typically aliphatic hydrocarbons) must not
only dissolve but somehow ionize....a rather tough challenge!
Typically, people rely on empirical remedies, but clearly don't
understand why they work. Dr. Croucher lamented that there have
been very few systematic, rational studies to propose and test
hypotheses of how charging takes place in dielectric fluids.
Reference was made to Dr. Fowkes of Lehigh University, who has
applied acid/base mechanisms to account for charged species in
composite hydrocarbon fluids. Without better understanding of
ionization in non-aqueous media and the effect of surfactants
at interfaces that give rise to charge, how does one go about
producing, much less controlling such charges?

The nature of the paper substrate is likewise a critical
component in the electrographic process. It is important that the
charge be localized on the paper surface; thus, the paper must be
dielectric on one side and conductive on the other. It is also
vitally important that paper coatings be consistent with respect
to uniformity, thickness, pin-holes (no conductive channels) and
overall quality control. Thus, the paper itself also requires
surfactants in its manufacture just as does the liquid developer.
The paper surface must also be compatible with the liquid developer.

The future direction of electrographic system development is
therefore highly dependent upon progress in understanding charging
processes in dielectric fluids. Dr. Croucher referred to a
latex-dye approach which he envisions will ultimately displace the
pigment-based formulation. Within Xerox, the largest proportion
of effort is devoted to electrographic developers rather than
ink jet, since it is more consistent with corporate culture.
Electrographic printing has found its niche in specialized applications; it offers such features as the ability to print in color
and to yield bigger print copies with thinner print layers than
typical dry xerography. The primary obstacle to preventing broader
application is the lack of understanding of charge development
at interfaces in nonaqueous solvents.

<u>Ink Jet</u>:   Currently, ink jet inks consist of multi-component
aqueous solutions. Although in theory deionized water-based
developers could also be used for electrographic printing, in
practice the water picks up ions so quickly that the imaging process is exceedingly difficult to control consistently. The transport process in ink jet devices obviates the need for charge control but amplifies the need for a minimum of 35 dyne/cm surface
tension to form the droplet for the jetting process. The latter
requirement basically excludes the use of non-aqueous solvents
and severely limits the applicability of such concepts as emulsions
and micro-emulsions. Water-based inks are likewise preferred
from an environmental standpoint because of minimal air pollution.
By comparison with electrography, ink jet systems are in general
less complicated, but problems still exist; e.g., unwanted precipitation, instability because of the large number of components and
changes observed as the amount of water evaporates or is absorbed.
Likewise, certain problems are encountered in the droplet formation and jetting processes themselves. In addition to assuring
colloid stability of the ink itself, the role of surfactants could
extend to paper surface preparation itself in order to broaden
the latitude of paper to be used with currently available ink
solutions.

   Several brainstorm ideas were proposed as possibilities to
consider: use of different surfactant aggregates, micro-emulsions,
inverse emulsions, micro-fluidization. Dr. Croucher volunteered
that he had demonstrated a vesicle approach for incorporation of
the dye. He noted that this approach basically worked, forming
prints with ability to jet nice droplets, but unfortunately the
system was not totally stable. He likewise mentioned a second
approach involving use of very small latex particles (200$\overset{\circ}{A}$), which
he has successfully dyed and jetted. In Croucher's opinion, there
are many opportunities for companies or individuals t o be innova-
tive in the area of ink jet printing; if anything, this application suffers from a lack of innovation. From a patent standpoint,
Croucher noted that he has filed numerous applications but a

rather recent generic patent by Fuji on latex particles for ink
jet inks has been a formidable obstacle for Xerox and other companies interested in ink jet development. Croucher noted that
there were not many patents for colloidal structures for ink jet
inks (10 at the outside, mostly Japanese). He also mentioned that
there was a good one on emulsions/micro-emulsions by a German
company. Dr. Croucher felt that since there was basically only
one type of multi-component jet ink solution, the bulk of which
being supplied from Japan, a fruitful approach would be to couple
the ink to paper so that image resolution could be optimized.
Other Imaging Technologies: The only other liquid developer system
mentioned was that pertinent to magnetic toners for MICR check
printing. Croucher mentioned that Xerox has invested very heavily
in liquid magnetic toner technology within the last two years but
such developers are currently only in the experimental stage.
With respect to magnetic particle needs for this application,
Croucher commented that the primary focus was on getting these
systems to work and that such needs would then be addressed once
the technology is mastered.

On the other hand, several dry developer systems were mentioned casually in passing. In general, liquid developers offer superior resolution relative to dry processes. Croucher pointed out
that the limitation to improving resolution via dry powders is the
mean size of the particles. Particle size for dry systems are
approximately 10 micron vs. <1 micron particles for liquid developers. Once below 5 micron size, dry particles easily become
airborne and quickly approach limits set by emission standards.
This problem can be alleviated somewhat through the use of magnetic scavengers but because most people are interested in high
resolution for most applications, liquid developers still have a
clear technical advantage. Croucher commented that in spite of
the greater complexity involved in dry developers, many man-years
of R & D investment have been expended by Xerox and other companies so that the dry development technology today is rather sophisticated and in reasonably good shape. In fact, Croucher admitted

that laser printing technology was probably in better shape than
electronic printing. Regarding the suggestion of magnetic transport as an alternate to electrostatic transport, Croucher commented that in theory such a technique could also work but he was
unaware that it had been attempted or demonstrated. An allied
printing option mentioned was magnetography, which today is a dry
process technology which has been thoroughly investigated at
Xerox and shown to be fully capable of making superb prints but
doubtlessly would never be commercialized by Xerox because of
direct competition with the 9000 series of laser printers/copiers
which Xerox is already marketing. Croucher elaborated further that
magnetography prints probably better than laser printers with fewer
unwanted background black spots.

2

Possible Applications of Surfactants in Microelectronics

Melvin Pomerantz

IBM, T. J. Watson Research Center
Yorktown Heights, New York

ABSTRACT

Surfactants may be deposited on solid substrates one molecular layer at a time. Such layers have special properties - thinness, uniform composition, variety of architectures, alignment of the molecules- which offer the possibility of novel applications. Some proposed applications in microelectronics, including very thin insulators, conductors, non-linear media, sensors, and magnetic media, are reviewed in this article.

I. INTRODUCTION

One of the most striking chapters in the history of technology is the rapid improvement in the capabilities of electronic circuitry. In the field of computer circuitry there has been a ten fold increase in the speed of logical operations in the past fifteen years [1]. A major part of this improvement has derived from scaling down the sizes of the transistors and associated components [2]. At present the state of the art involves dimensions of the order of about one micrometer; further diminution of the size is being vigorously pursued in laboratories around the world. The size of a circuit element currently is about as small as a living cell. Soon it will be comparable to a large macro-molecule. It is thus not too soon to start thinking about circuit elements on a molecular scale. Surfactant molecules have properties that seem to offer possibilites to aid in the fabrication of circuit elements in the submicron range. Furthermore, by depositing surfactants one monolayer at a time, materials may be constructed that have desired functions. Such applications of surfactants will be the subject of this review. There are other uses of surfactants in the preparation of magnetic recording media and ce-ramics. These are important applications in the electronics industry, but not in the sense of "microelectronic" as I have used it. I shall not include these because they will be reviewed in other articles in this volume.

This review is limited to those applications in which the surfactant is applied in single monolayers at a time. (A rather more extensive survey has been given by Roberts [3].) In the next Section (II) I describe and then compare the three techniques of monomolecular deposition: the Langmuir-Blodgett (LB) method, the self-organizing method, and thermal evaporation. Section III describes some of the meth-ods by which monolayers and multilayers are observed and characterized. Part IV covers some potential applications in which the surfactant acts as a "passive" partic-ipant in some process. Examples are very thin insulating layers and capacitors. Part V is a review of some proposed applications in which the surfactant is "active" in transforming an input signal into an output that is different. Sensors and nonlinear devices are examples of this type. In concluding, I indicate some of the research that will be necessary to achieve these applications.

II. METHODS OF DEPOSITION OF SURFACTANT MONOLAYERS

The microelectronic applications considered here are those in which the surfactant is deposited as a single monolayer or as multilayers in a controlled and reproducible manner. There are three methods currently being employed to achieve this.

A. THE LANGMUIR-BLODGETT TECHNIQUE

I shall present only the basic idea of the method because there is a book [4], conference proceedings [5,6], a special issue of a journal[7], and an extensive review article [8] in which details are available. The Langmuir-Blodgett method is based on the tendency of oils and water to separate when mixed, which is the basis of surfactant behavior. The molecules used in the Langmuir-Blodgett method are surfactants ; one of their ends is hydrophobic and the other is hydrophilic, and they are insoluble in the subphase (usually water). Thus, the molecules tend to accumulate on the surface of water. They align with the hydrophilic end submerged, and the hydrophobic end as much out of the water as possible. This orientation is enhanced when surface pressure is applied. A paradigm for the surfactants used in the Langmuir-Blodgett method is the class of long-chain carboxylic acids. An example of this is $C_{17}H_{35}COOH$, octadecanoic acid, commonly known as stearic acid. Its structure is illustrated in Fig.1a. In the Langmuir-Blodgett method a small amount of such molecules, dissolved in a volatile solvent, is spread on the surface of water. The solvent evaporates, leaving a "gas" of surfactant molecules on the surface, as illustrated in Fig. 1b. These molecules are compressed by a movable surface piston until a relatively incompressible solid film is formed (c). This pressure is maintained while a suitable substrate is inserted and removed through the surface. When all goes well, the film is transferred from the water surface and adheres to the substrate. Thus a single monolayer may be deposited, as illustrated in Fig.1e. If the substrate is reinserted another layer attaches (f), and another upon exiting (g). The usual way in which the layers attach to each other and to the substrate is that the hydrophobic surfaces attach to each other and the hydrophilic surfaces attach to each other. Other structures are possible.

B. SELF-ORGANIZING FILMS

This method utilizes the chemisorption of surfactant-type molecules onto a substrate. The substrate is simply immersed in a solution of the surfactant in an appropriate solvent. A typical surfactant used by Sagiv and collaborators [9] is a bifunctional silane of the general form $CH_2 = CH - (CH_2)_n -SiCl_3$ which is dissolved in 80% n-hexadecane-12% $CCl_4 - 8\% CHCl_3$. The silane group readily bonds to exposed free hydroxyl groups on the substrate surface. The terminal vinylic group is not bulky, thus allowing closely packed layers. The reason that this group is incorporated in the molecule is that it can be chemically transformed into a hydroxyl group. This permits a second silane layer to bond. Thus, a layer-by-layer buildup can be achieved. As with

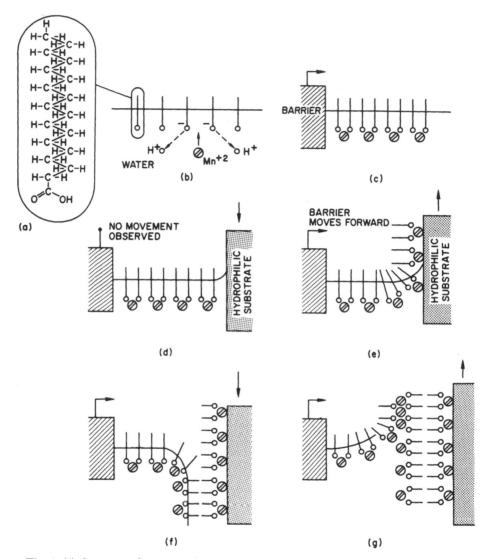

Fig. 1 (a) Structure of stearic acid, a typical surfactant used in the Langmuir-Blodgett technique. (b) - (g), the steps of the Langmuir-Blodgett deposition. In (b), stearic acid dissolved in a volatile solvent, is spread on the water. The deposition is often facilitated by the conversion of the acid to its salt. This may be done by dissolving divalent ions in the water. The example shown here is Mn^{+2} ions. At an appropriate pH (7, in this case) the acid is ionized and reacts with Mn^{+2} to produce manganese stearate. (c) The monolayer is compacted. (d) A hydrophilic substrate (e.g., glass) is inserted. The barrier is stationary indicating that no film is removed from the water. If the substrate were hydrophobic, (e.g. graphite) a layer would deposit, with the hydrophobic end ("tail") toward the substrate. (e) Lifting the substrate, film is deposited. The barrier moves forward. (f) Second layer is deposited by reinserting the substrate. Layers attach tail-to-tail. The meniscus turns downward and barrier moves forward. (g) Third layer is deposited on lifting the substrate. Many layers can be built up by repeating.

the Langmuir-Blodgett method, the successive layers need not be identical. It was recently confirmed that in some cases the coverage by straight chained silanes approaches closely to 100% for a single monolayer [10]. For unsaturated chains or
multilayers the layers appear to be less than complete. The completeness depends on
the structure of the molecules, which determines the packing, and the conditions of
preparation.

C. THERMAL EVAPORATION

There are several reports of the deposition of monolayers of surfactants by thermal
evaporation of molecules such as stearic acid [11, 12]. One advantage of this method is
that during the deposition the sample is not immersed in a liquid. Furthermore, evaporation is more compatible with the other steps in the processing of integrated circuit
devices. The completeness of coverage of the substrates for thin layers is not as good
as Langmuir-Blodgett films. However, by applying voltages to the films, a healing is
achieved which increases the electrical resistance and the electrical breakdown strength
[13]. The evaporation method is appropriate for producing films of a single type of
molecule but it does not seem to offer the possibility of building a precise pattern with
a variety of molecules.

The three methods described above differ in their ease of operation as well as the
kinds of surfaces they can cover. The Langmuir-Blodgett method probably is the
fastest in changing from one kind of molecular layer to another. One need only have
the desired films spread on separated water surfaces, and pass the substrate from one
to the other. The self-organizing method, by contrast, requires a chemical step to ac-
tivate the hydrophobic end. The solution above the substrate must be changed and the
surface reacted. The evaporation method cannot produce a monolayer reliably, so
precise molecular layering seems impractical.

A characteristic feature of the Langmuir-Blodgett method is that for reasonably
smooth surfaces the amount of film transferred from the water surface is often about
equal to the projected geometrical area of the substrate [14, 15, 16]. This is interpreted
to mean that the film does not follow the atomic topography; it appears to bridge over
the roughness of the substrate. Such a condition implies that the Langmuir-Blodgett
film is smoother than the original substrate. Indeed, such a smoothing effect was no-
ticed in the decrease in the apparent roughness needed to explain x-ray diffraction in

multilayers compared to a monolayer [17]. This property can be a disadvantage if the goal is to coat the entire surface of a substrate. The Langmuir-Blodgett method fails entirely for highly irregular substrates, such as a powder. In order to coat all the surfaces of a powder the self- organizing approach is uniquely appropriate. It is interesting that the self-organizing film may also bridge across small asperities on a surface. One would not expect this at first glance, because it would seem that the molecules would plate out of the solution and cover all the surface. However, the x-ray diffraction from self organizing films on Si wafers were best explained by a dramatic increase in smoothness of the film compared to the original surface [10]. The likely explanation derives from the fact that the silane ends of the molecules polymerize with each other. The sidewards bonding perhaps confers strength on the film, so that it can suspend across the depressions in the substrates, rather than coating them.

III. CHARACTERIZATION OF MONOLAYERS AND MULTILAYERS

It is not sufficient to construct monolayer structures; it is also necessary to be able to obtain measureable signals from them. The possibilities may be inferred from a review of the means by which such films are characterized. It is remarkable that quite a lot can be learned about even a single monolayer, a quantity of only about a microgram/ cm^2, by a variety of sensitive techniques. For example, it is important to verify that the layer structure we believe we produce by the monolayer techniques is actually present on the substrate. Detailed confirmation was obtained [18, 17] by *x-ray* diffraction. Normally X rays are very penetrating and one would not expect to observe a diffraction effect from a single molecular layer. The Langmuir-Blodgett film, however, has special advantages: its thickness of about 2.5 nm, and a unit cell of two molecules (hence 5.0 nm), results in a Bragg angle of about 1° from the grazing angle. This is close to the angle for total internal reflection of X rays and thus the intensity of the interfering beams is relatively large. Fig. 2 illustrates the data we obtained from 1, 3 and 5 layer films on 4 cm^2 Si substrates. This required a diffractometer with excellent collimation and monochromaticity, and counting times of about 5 minutes per point. The structure is sufficiently simple that it can be modeled as laminae of the known chemical constit-

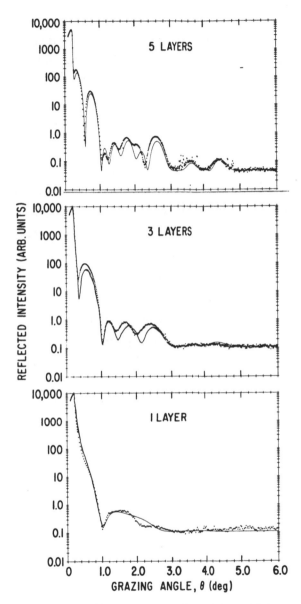

Fig. 2 *X-ray diffraction patterns from 1, 3, and 5 layers of manganese stearate on Si substrates. The dots are experimental points, the solid curve is calculated from a model structure* [17].

uents of the molecule [17]. The resulting calculated diffraction pattern, shown by the solid lines in Fig. 2, is in excellent quantitative agreement with the data. The detailed pattern of major peaks and subsidiaries is directly related to the number of layers in the film.

It has proved possible to measure the the position of particular atoms in crystal structures by the use of *x-ray standing waves* [19]. This technique was applied to Langmuir-Blodgett films by Nakagiri, et al. [20]. Using synchrotron radiation, they measured the position of a single layer of Mn ions in Mn stearate to a resolution of ± 0.5Å. They estimated that as little as 0.01 of a monolayer would be detectable under realizable conditions. It has also been possible to observe [21] *neutron* diffraction from a small number of Langmuir-Blodgett layers, which may be useful for studying effects of deuteration and magnetic effects, both of which produce distinctive effects on neutron diffraction. The information obtained relates to the structure perpendicular to the planes.

The determination of the in-plane structure of Langmuir-Blodgett films was initiated soon after the technique was developed, with the use of *electron* diffraction [22]. Very few layers are required. For example, the structure of as few as two monolayers of manganese stearate was observed [23] by S. Herd and the author. The structure of manganese stearate turned out to be similar to that of Pb stearate, which had been measured earlier by electron diffraction [24]. The implication of this similarity is that the structure is not much affected by even drastic changes in the size of the metal ion. Yet, although the structure is mostly determined by the alkane chains, it is not simply hexagonal close packed, as one might first guess. For more complicated molecules, such as an anthracene derivative [25], as expected, the in-plane structure is found to be of low symmetry.

The use of high energy electrons in electron diffraction has the effect of rapidly degrading the film. A less destructive technique employs the *non-collinear reflection-diffraction of X rays*. The in-plane structure of Langmuir-Blodgett films of Pb stearate was measured by this method [26]. The x-ray scattering was maximized in these pioneering measurements by having some hundreds of layers and by using a heavy metal ion (Pb). As techniques improve more general applicability can be expected. Measurements of the inter-chain spacing have also been made by *low energy electron diffraction* (LEED)[27]. Values of these spacings have been obtained recently from images of the monolayers produced with a *scanning tunneling microscope* [28].

Structural information about films can also be obtained from *polarized infra-red absorption*. E.g., one utilizes the property that the dipole moment of the C - H bond on an alkane chain is perpendicular to the chain length. If infra-red radiation that has a polarization perpendicular to the substrate is absorbed it means that the chains are not perpendicular to the substrate. Such effects can give information about motion of the chains, e. g., melting has been studied [29]. Total internal reflection, combined with Fourier transform methods, provides sensitivity on the monolayer level of coverage.

For some applications it is necessary to convert an acid surfactant to its salt, a soap. Such a reaction can be monitored by *IR absorption* . Thus, in carboxylic acids the vibration of $O = C O H$ is changed upon reaction to COO^-M^+, where M^+ is a metal ion [30] . This provides a convenient and sensitive measure of the attachment of the metal ion to this specific site. For example, in some of our experiments we wished to make the monolayers magnetic. We bonded magnetic ions (Mn^{+2}) to the acidic ends of the molecules. Following some results in the literature [31] and our experiments [23] we determined that Mn^{+2} ions would replace the H^+ when the concentration of $MnCl_2$ was 10^{-3} molar and the pH was approximately 7. Under these conditions a monolayer of Mn stearate, $Mn(C_{18}H_{35}OO)_2$, is formed at the surface. The completeness of the reaction was monitored by skimming the film off the surface on to a Si substrate. The infra-red spectra were recorded on a simple Beckman spectrograph. We could deduce the bonding of the Mn to the molecule by the reduction in the $O=COH$ carbonyl peak at $6\mu m$ and the growth of an absorption at 6.5 μm, characteristic of bonded metal ions [30]. Using a more sophisticated surface-wave technique it was possible to observe IR absorption by as few as one monolayer [32].

It is sometimes necessary to establish the quantity of a particular element in a sample containing only few monolayers. Analytical techniques that have the required sensitivity include Auger electron spectroscopy, ESCA, Rutherford backscatering, and electron microprobe (i.e., excitation of characteristic photon emission by an electron beam).

In cases in which the films contains paramagnetic entities it may be possible to observe *electron spin resonance (ESR)*. The anisotropy of the resonances has been used to deduce the orientation of the magnetic species with respect to the plane and the direction in which the film was pulled [33, 34, 35]. The anisotropies due to the dipolar interaction in a two-dimensional array [36] have also been observed in Langmuir-Blodgett film of manganese stearate [18]. The Langmuir-Blodgett technique is a natural means for producing precisely one layer of magnetic ions, i.e., a two-dimensional magnet. The prediction was that if the magnetic moments were in a two-dimensional array, the ESR should possess anisotropy in both the line width, ΔH, and line position, H_0, as a function of θ, the angle of the external magnetic field with respect to the film normal. The line width effect arises because the dipolar broadening depends on spin diffusion, and diffusion depends on dimensionality. In two dimensions there are long correlation times which lead to enhancement of the term in the dipole interaction that is at zero frequency, namely the term in ΔH proportional to $(3 \cos^2\theta - 1)$. This was observed to be the dominant contribution. The dipolar interaction also contributes an average magnetic field that is anisotropic in two-dimensions. This results in a shift of H_0 to higher values in the perpendicular direction (to overcome the opposing dipole fields), and to lower fields in the in-plane orientation. This is a small effect but was observed in a multilayer film in the highest field available to us. The significance of these ESR results in the paramagnetic phase is that they demonstrate the anisotropic magnetic characteristics expected for two-dimensional arrays. It proves that there is not a significant amount of Mn in three-dimensional environments, as might occur if a significant amount of water containing dissolved Mn were drawn along with the films. The magnetism of a few monolayers of manganese stearate was also observed using *squid magnetometry* [37]. These measurements indicated a magnetic ordering transition at about $0.3\,^\circ K$.

IV. APPLICATIONS OF THE "PASSIVE" TYPE

"Passive" behavior of the surfactant film is defined here by the output from the film being equal to the input to it. A simple but important example is an *electrically insulating barrier*. The need arises in the fabrication of field effect transistors (FET) to electrically isolate a gate electrode from the channel region. One of the major reasons for the dominance of silicon for FET devices is the fact that an insulating barrier is readily formed on its surface by oxidation. For those semiconductors that do not have adequate native oxides it is necessary to find a thin continuous insulating barrier. There have been a number of reports [38, 5] of the successful application of Langmuir-Blodgett films on semiconductors such as CdTe, InP, and GaAs. For example, a GaAs device, employing as an insulator a three layer Langmuir-Blodgett film, showed sharp switching characteristics [39]. The surfactant used in this case was ω − tricosenoic acid, a 23 carbon chain with a terminal vinyl group. One possible advantage of using surfactants as insulating layers is that they are deposited at about room temperature; such mild conditions might not affect the semiconductor. In fact, some influence on Schottky barrier heights has been reported [40]. The origin of the effect was suggested to be the dipole moment of the ionic end of the molecule. Other applications of very thin insulating surfactants include capacitors [41] and tunnel barriers [42, 43].

A difficulty with the utilization of simple long-chain molecules is that their melting points are less than 100°C. Thus, any processing that follows their deposition must be at temperatures lower than about 100°C in order to avoid damaging the insulator. This rather severe restriction may be overcome by using more robust molecules. Progress in this direction has been made by the development [44] of a Si-phthalocyanine molecule which can be deposited sufficiently continuously to support quite large electric fields, with a resistivity of about 5×10^5 Ωcm. Similar phthalocyanine molecules are thermally stable to about 300 C; clearly this class is more suitable for postprocessing. Another approach to the problem of film instability is to use molecules that are polymerizable [45].

In contrast to insulation, it may be desirable to have monolayers that are *electrically conducting*. There has been work on films that become conducting because of doping or irradiation, and recently films have been developed whose conductivity is intrinsic. In the former group are Langmuir-Blodgett films of N-docosylpyridinium-TCNQ complex. They become conducting upon exposure to iodine vapor [46, 47]. The conductivity is about 0.1 $(\Omega\text{cm})^{-1}$. Intrinsic conductivities of similar magnitude have been achieved in Langmuir-Blodgett films of TMTTF-octadecylTCNQ, by Nakamura, et al [48]. The sensitivity of these organic layers to chemical and physical changes opens the possibility of local modification of the conductivity. Namely, there is the possibility of writing conducting wires in an insulating film.

In the experiments reported thus far it is not sure that a perfectly continuous film has consistently been achieved with a single layer. Improvements in techniques and/or materials are needed.

In addition to electrical applications, there may be uses for surfactant films that exploit their rheological or structural properties. They are known to be lubricants [49]; an application to lubricate and protect metal magnetic tape has been described [50]. By modifying their chemical structure it may be possible to create an adhesive. Considerable improvement of abrasion resistance is needed to make these practical. The orderly array of surfactant molecules makes them suitable as spacers on a nanometer scale. One might envision an application in which the fluorescent lifetime of a molecule is adjusted by spacing it above a conducting surface [51, 8, 52]. There have been proposals to synthesize molecules that can rectify electric current [53], like a p-n junction, or can act as switches [54]. To be useful these molecules would have to be oriented in a uniform way, say with all the n sides away from a particular electrode. A natural way to achieve this molecular alignment is to add appropriate groups to the molecules to cause them to become a surfactant , and thus to orient on the surface of water. The molecules could then be deposited by the Langmuir-Blodgett method.

V. SURFACTANTS IN "ACTIVE" APPLICATIONS

By an "active" application I mean a case in which the output from the surfactant layer is different from the input to it. There is at present a commercial application that illustrates an active function, although it is not electronic. Langmuir-Blodgett films of Pb carboxylates are used as diffraction gratings for soft X rays [55]. The emerging X rays are changed in direction, particularly intensely at the Bragg angle.

An electrical application that beautifully utilizes the Langmuir-Blodgett method is the development of a temperature *sensor* based on the pyroelectric effect [56, 57]. Pyroelectricity is the change in the electric potential across a material when its temperature changes. In order for this to occur the material must have a spontaneous and permanent electric dipole moment: it must be ferroelectric. This in turn means that the centers of + and - charge in the material must be separated. If the temperature changes, the lattice expansion or changes in the charge density on the surface cause the potential across the material to change. The Langmuir-Blodgett method is well suited for the construction of a material with a permanent dipole moment. This may occur, to some extent, whenever different films are deposited alternately. The dipole moment can be maximized by alternating a surfactant that is anionic with one that is cationic. Christie, et al, [58] have used ω −tricosenoic acid (anionic) alternating with docosylamine (cationic). They found that a 99-layer sample has a pyroelectric activity comparable to that of polyvinylidene fluoride, a commercially used pyroelectric.

In a similar way, the structural properties needed for *non-linear optical processes* can be constructed. The symmetry requirement for second harmonic generation is that the material not have a center of symmetry. This is the same condition as for piezoelectricity. Langmuir-Blodgett films with an odd number of layers will generally lack the center of symmetry, and thus be suitable. A judicious choice of alternant layers can lead to improvement of optical harmonic generation [59, 60, 61] or ultrasonic transduction via the piezoelectric effect.

An interesting application that utilizes molecular architectural design is the construction of a p-n junction *rectifier* by the combination of films which separately have p and n character. E.g., Sakai, et al [62] used paraquat, with two stearyl chains attached to make it a surfactant , as the n-type layer. For the p-type layer, one of several merocyanine dyes with substituted stearyl chains was chosen. These were deposited alternately by the Langmuir-Blodgett technique, such that the p and n layers were adjacent to each other. It was found that in the dark the current-voltage characteristic was rectifying. The magnitude of the current depended on the wavelength of the illumination, in proportion to the absorption by the materials. The non-linearity of the I-V curves was not that of a convential p-n diode (it varied as $\log I \propto V^{1/2}$, which is characteristic of bulk conduction in Langmuir-Blodgett films). The details of the conduction mechanisms need clarification [63].

Another concept, advanced by Kuhn and collaborators [64, 65], is a suitably designed molecular structure which would *generate electricity from sunlight*. The successive layers would serve as antenna, transporter of any generated electric charge, perhaps a layer to store the charge, and then means to utilize the charge. Further impetus to pursue this idea may have to await another oil crisis.

In addition to these electrical applications there has been work on *magnetic* surfactant films. As remarked in Section III above, magnetic ions of Mn can be chemically bonded to the ionic end of the molecule. Layers of such films can be deposited on substrates by the Langmuir-Blodgett method. Experiments [23] have been done on the magnetism of single atomic sheets of these Mn ions, on plates stacked up to increase the total area to about 200 cm². The important properties for applications of magnetic films are the temperature at which spontaneous ordering occurs, and the kind of ordering that occurs. The Mn stearate Langmuir-Blodgett films proved to have an ordering temperature no higher than 2°K - much too low to be practical. The kind of magnetic order that appeared in the Mn stearate produced only a weak magnetic moment, which also is not well suited for applications. In another sense, however, the result was encouraging because there were theoretical predictions [66] which suggested that the monolayer magnet might not be able to order magnetically at any non-zero temperature. In fact, however, it is unlikely that any realistic magnet would satisfy the assumptions of the theory. The experiments with the magnetic surfactant offer evidence that even a material as thin as a single magnetic atom is not prohibited from being spontaneously magnetically ordered. The practical advantages of such a thin magnet, if it were ferromagnetic at room temperature, are that it should have faster switching speed and higher storage density than thicker films. Clearly much improvement is needed if such surfactant magnets are to be realized.

Thus far I have discussed applications in which the thinness of the surfactant film played a helpful role. Thinness and uniformity may also confer advantages for use as a *resist medium* in circuit fabrication. In microelectronics it is the lateral dimensions that determine how densely the circuit elements can be packed onto a chip. Optical lithography is used currently to imprint the desired patterns in a photosensitive resist layer. The fundamental limits of this method are currently being approached. The limit is set by diffraction of the light as it passes through the masks that define the pattern being produced. It is possible to resolve lines only if they are separated by more than about one wavelength of light. Thus optical methods will be limited to separations of about 0.5 μm. To make yet closer spacing, shorter wavelength radiation (U V and X rays) are being considered. Another approach is to use electron beams, which can have beam diameters of order 5 Å or less. But not only will the radiation dimensions

need to be reduced, also the resist medium will need refinement. At present the resist
is spun onto a substrate to a thickness of the order or less than the intended line spacing. When the spin coating is less than 1000 Å thick there may be problems because
patches of the substrate may be uncoated and therefore some part of the circuit will
be incorrectly patterned. A few attempts to solve this problem using Langmuir-
Blodgett films have been reported. The obvious advantages of the Langmuir-Blodgett
process are that films are of quite uniform thicknesses and excellent coverage. The
disadvantages are that the deposition of, say, 50 layers is time-consuming (at least an
hour for a 5 " wafer), and the chip must be submerged in water, which it may resent.
The experimental results thus far support the intuition that high resolution is possible.
On a thin substrate (200 Å of amorphous C) coated with 100 layers of Mn stearate,
an electron beam drew lines that were resolved at 100 Å separation [67]. The thinness
of the substrate is a factor because the lines drawn by the fine electron beam may be
broadened by the backscattering of the electrons from the substrate. With a thin
substrate the electrons pass through without backscattering. More realistic experiments on thick aluminum substrates, using both ω −tricosenoic acid layers and other
polymerized films, were reviewed and discussed by Barraud [68] . Here lines separated
by as little as 900 Å were resolved. In the meantime, skillful workers have improved
the spin-on technique such that lines separated by less than 1000 Å have been achieved
by this technique. It seems likely, however, that the Langmuir-Blodgett technique will
eventually prove to have superior performance when the thicknesses are even less.
With fewer layers it will also take less time to deposit the coating.

VI. CONCLUSIONS

It is clear from even this incomplete survey of activities that surfactants have remarkable properties that might prove to be useful as electronic circuit elements approach
molecular sizes. One would benefit from the thinness of the films in order to make
compact layers that either inhibit, conduct or even rectify electric currents. It is known
how to create artificial structures into which desired functions may be built. The
pyroelectric sensor is a paradigm for this kind of invention. The application as a resist

medium is highly promising. There are, however, some difficulties that have impeded progress in each of the proposed applications. In some cases it is that the films thus far produced have not given perfect coverage of the substrates. (The search for the universal insulating barrier, that can be deposited at room temperature, is not ended.) The commonly used materials, typically long chain organics, are soft and fragile. There is a need for materials that can withstand thermal and mechanical stresses better.

The needs for improved materials, combined with measurements of their physical properties, points to an interdisciplinary team dedicated to this field. Such a realization has arisen in a number of places around the world. In England, a national program has been organized. In Japan the study of Langmuir-Blodgett and related techniques has been targeted as an area of intense development because of its promise in both technology and science. It seems very likely that significant results will be achieved.

REFERENCES

1.   M. S. Pittler, D. M. Powers, D. L. Schnabel, *IBM J. of Res. and Dev. 26:* 1 - 11 (1982).

2.   R. W. Keyes, *IEEE J. of Solid-State Circ. SC-14:* 193-201 (1979).

3.   G. G. Roberts, *Adv. In Physics 34:* 475-512 (1985).

4.   G. L. Gaines, *Insoluble Monolayers at Liquid-Gas Interfaces*, (Interscience, New York, 1966).

5.   G. L. Gaines (Ed.), *Proceedings of the Second International Conference on Langmuir-Blodgett Films, Thin Solid Films 132-134:* 1 (1985).

6.   G. G. Roberts and C. W. Pitt (Eds.), *Proceedings of the First International Conference on Langmuir-Blodgett Films, Thin Solid Films 99:* 1 - 329 (1983).

7.   W. A. Barlow (Ed.), *Special Issue on Langmuir-Blodgett Films, Thin Solid Films 68:* 1 - 288 (1980).

8.   H. Kuhn, D. Möbius, and H. Bücher, in *Physical Methods in Chemistry I*, edited by Weissberger and Rossiter, (Wiley, N. Y., 1972), Chap. VII, pp. 577 - 701.

9.   L. Netzer, R. Iscovici, and J. Sagiv, *Thin Solid Films 100:* 67 (1983).

10. M. Pomerantz, A. Segmüller, L. Netzer, and J. Sagiv, *Thin Solid Films 132:* 153-162 (1985).

11.   M. A. Baker, *Thin Solid Films 8:* R13 - R15 (1971).

12.   V. K. Agarwal, Y. Igasaki, and H. Mitsuhashi, *Jap. J. of App. Phys. 15:* 2327-2332 (1976).

13.   V. K. Agarwal and C. H. Huang, *Thin Solid Films 137:* 153-160 (1986).

14.   J. A. Spink, *J. Coll. and Interface Sci. 23:* 9-26 (1967).

15.   E. P. Honig, J. H. Th. Hengst, and D. den Engelsen, *J. Coll. and Interface Science 45:* 92-102 (1973).

16.   W. J. Plieth and W. Höpfner, *Thin Solid Films 28:* 351-356 (1975).

17.   M. Pomerantz and A. Segmüller, *Thin Solid Films 68:* 38 (1980).

18.   M. Pomerantz, F. Dacol, and A. Segmüller, *Phys. Rev. Lett. 40:* 246 (1978).

19.   P. L. Cowan, J. A. Golovchenko, and M. F. Robbins, *Phys. Rev. Lett. 44:* 1680 - 1683 (1980).

20.   T. Nakagiri, K. Sakai, A. Iida, T. Ishikawa, and T. Matsushita, *Thin Solid Films 133:* 219-225 (1985).

21.   R. M. Nicklow, M. Pomerantz, and A. Segmüller, *Phys. Rev. B23:* 1081 (1981).

22.   L. H. Germer and K. H. Storks, *J. of Chem. Phys. 6:* 280 (1938).

23.   M. Pomerantz, in *Phase Transitions in Surface Films*, edited by J. G. Dash and J. Ruvalds, (Plenum Press, New York, 1980), p. 317.

24.   J. F. Stephens and C. Tuck-Lee, *J. Appl. Cryst. 2:* 1 (1969).

25.   I. R. Peterson, G. J. Russell, D. B. Neal, M. C. Petty, G. G. Roberts, T. Ginay, and R. A. Hann, *Phil. Mag. B 54:* 71-79 (1986).

26.   M. Prakash, P. Dutta, J. P. Ketterson, and B. M. Abraham, *Chem. Phys. Lett. 111:* 395 (1984).

27.   V. Vogel and C. Woll, *J. Chem. Phys. 84:* 5200 (1986).

28.   J. Rabe, Ch. Gerber, J. D. Swalen, D. P. E. Smith, A. Bryant, and C. Quate, *Bull. Am. Phys. Soc. 31:* 289 (1986).

29.   C. Naselli, J. F. Rabolt, and J. D. Swalen, *J. Chem. Phys. 82:* 2136 (1985).

30.   B. Ellis and H. Pyszora, *Nature 181:* 181 (1958).

31.   G. A. Wolstenholme and J. H. Schulman, *Proc. Farad. Soc. 46:* 475 (1950).

32.   A. Hjortsberg, W. P. Chen, E. Burstein, and M. Pomerantz, *Optics Comm.*
       25: 65 (1978).

33.   J. Messier and G. Marc, *J. de Physique 32:* 792 - 804 (1971).

34.   P.-A. Chollet, *J. of Phys. C 7:* 4127 - 4134 (1974).

35.   S. Kuroda, K. Ikegami, M. Sugi, and S. Iizima, *Sol. State Comm. 58:* 493 - 497
       (1986).

36.   P. M. Richards, in *Proceedings of the International School of Physics "Enrico
       Fermi", Course LIX,* edited by K. A. Müller and A. Rigamonti, (North Holland,
       Amsterdam, 1976), p. 539.

37.   T. Haseda, H. Yamakawa, M. Ishizuka, Y. Okuda, T. Kubota, M. Hata, and K.
       Amaya, *Solid State Comm. 24:* 599 (1977).

38.   G. G. Roberts, in *Insulating Films on Semiconductors,* edited by M. Schulz and
       G. Pensl, (Springer Verlag, Heidelberg, 1981), p. 56.

39.   N. J. Thomas, M. C. Petty, G. G. Roberts, and Y. H. Hall, *Electronics Lett.
       20:* 838 (1984).

40.   C. S. Winter and R. H. Tredgold, *J. of Phys. D. 17:* L123 - L126 (1984).

41.   H. Nakahama, S. Miyata, T. T. Wang, and S. Tasaka, *Thin Solid Films 141:* 165
       - 169 (1986).

42.   B. Mann and H. Kuhn, *J. App. Phys. 42:* 4398 (1971).

43.   S. Hao, B. H. Blott, and D. Melville, *Thin Solid Films 132:* 63 - 68 (1985).

44.   Y. L. Hua, G. G. Roberts, M. M. Ahmad, M. C. Petty, M. Hanack, and M. Rein,
       Phil. Mag. B 53: 105 - 113 (1986).

45.   A. S. Dewa, C. D. Gung, E. P. Dipoto, and S. E. Rickert, *Thin Solid Films
       132:* 27 -32 (1985).

46.   A. Ruaudel-Teixier, M. Vandevyver, and A. Barraud, *Mol. Cryst. Liq. Cryst.
       120:* 319 (1985).

47.   A. Barraud, A. Ruaudel-Teixier, M. Vandevyver, and P. Lesieur, *Nouv. J. Chim.
       9:* 365 (1985).

48.   T. Nakamura, M. Matsumoto, F. Takei, M. Tanaka, T. Sekiguchi, E. Manda,
       and Y. Kawabata, *Chem. Lett. (Jap.) 5:* 709 - 712 (1986).

49.   B. J. Briscoe, D. C. B. Evans, and D. Tabor, *J. of Colloid and Interface Science
       61:* 9 - 13 (1977).

50.   J. Seto, T. Nagai, C. Isimoto, and H. Watanabe, *Thin Solid Films 134:* 101 -
       108 (1985).

51.   K. H. Drexhage, *J. of Luminescence 1,2:* 693 - 701 (1970).

52.   A. C. Pineda and D. Ronis, *J. Chem. Phys. 83:* 5330-5337 (1985).

53.   A. Aviram and M. Ratner, *Chem. Phys. Lett. 29:* 277 - 283 (1974).

54.   A. Aviram, P. E. Seiden, and M. Ratner, in *Molecular Electronic Devices,* edited by F. Carter, (M. Dekker, N. Y., 1982), Chap. I, pp. 5 - 17.

55.   M. W. Charles, *J. App. Phys. 42:* 3329 (1971).

56.   L. M. Blinov, N. V. Dubinin, L. V. Mikhenev, and S. G. Yudin, *Thin Solid Films 120:* 161 (1984).

57.   G. W. Smith, M. F. Daniel, J. W. Barton, and N. Ratcliffe, *Thin Solid Films 132:* 125 - 134 (1985).

58.   P. Christie, G. G. Roberts, and M. C. Petty, *App. Phys. Lett. 48:* 1101-1103 (1986).

59.   P. A. Chollet, F. Kajzar, and J. Messier, *Thin Solid Films 132:* 1 -10 (1985).

60.   F. Kajzar and J. Messier, *Thin Solid Films 132:* 11 - 20 (1985).

61.   I. R. Girtling, N. A. Cade, P. V. Kolinsky, J. D. Earls, G. H. Cross, and I. R. Peterson, *Thin Solid Films 132:* 101 -112 (1985).

62.   K. Sakai, M. Saito, M. Sugi, and S. Iizima, *Jpn. J. of App. Phys. 24:* 865 - 869 (1985).

63.   P. S. Vincett and G. G. Roberts, *Thin Solid Films 68:* 135 - 171 (1980).

64.   H. Bücher, K. H. Drexhage, M. Fleck, H. Kuhn, D. Möbius, F. P. Schäfer, J. Sondermann, W. Sperling, P. Tillman, and J. Wiegand, *Molecular Crystals 2:* 199 - 230 (1967).

65.   H. Kuhn, *Pure and Applied Chemistry 53:* 2105 - 2122 (1981).

66.   N. D. Mermin and H. Wagner, *Phys. Rev. Lett. 17:* 1133 (1966).

67.   A. N. Broers and M. Pomerantz, *Thin Solid Films 99:* 323 - 329 (1983).

68.   A. Barraud, *Thin Solid Films 99:* 317 - 321 (1983).

Microelectronics
Group Discussion I

Richard M. Mullins, Olin Chemicals Group, rapporteur

The discussion was limited to applications of Langmuir-Blodgett (L-B) films in two areas, specifically, thin resists and chemical sensors.

Current technology for the preparation of resists utilizes the spinning of polymer films to produce resists. These films are probably thicker (1 micron) than they need to be. It is conceivable that L-B films could produce thin films (ca 500A$^\circ$) of uniform thickness. In order to achieve this, the surfactant molecules would have to be multifunctional, enabling them to be crosslinked into continuous polymeric films via some means (e.g., electron beams, X-ray). This technology might improve the quality of resists because of the uniformity of the film thickness. In addition, the method might allow further size reduction in the electronics industry.

Another application for L-B films is in chemical sensors (for pollutants, toxic gases, etc.). Current technology can utilize L-B films to prepare workable films, but the films are not permanent, i.e., they are subject to degradation from abrasion, temperature effects, and other adverse environmental factors. Crosslinkable L-B films could provide stronger films for chemical sensors and still retain the necessary thinness and uniformity required for fast response.

The general trend seen was the need for surfactant molecules that can be placed where they are needed (for example, using L-B film techniques) and then crosslinked to make them more permanent.

Microelectronics
Group Discussion II

Robert Falk, CIBA-Geigy Corp., rapporteur

Microelectronic applications include insulators, conductors, capacitors, rectifiers, fluorescers, and chemical sensors. The objective here is for the preparation of circuit boards resists that can be grooved by X-ray or electron beams. Photolithographic resolution of 0.5-1.0 microns is limiting current technology.

The discussion centered on the production of two-dimensional magnets, using molecular films of Mn salts. Magnetic salts of other metals could also be used. Preparation of these films by the Langmuir-Blodgett technique was discussed. They could be anchored by chemisorption onto silicon. At present, magnetic effects of these films at $2^\circ K$ have been demonstrated, but activity at higher temperatures is desired. Metallic salts of fluorocarbon acids was suggested for possible use in this application.

3
The Role of the Surfactant in Magnetic Recording Media

Mark S. Chagnon and Robert Donadio

Integrated Magnetics, Inc.
Chelmsford, Massachusetts

Introduction

Surface activity has been defined as the pronounced tendency of a solute to concentrate at an interface. Molecules having certain chemical structures exhibit surface activity. In general, a surface active agent can be defined as any compound that has a hydrophilic head group and a hydrophobic tail. Soaps (salts of fatty acids), long chain amino acids, and lipids all fall into this class of compound. In fact, any molecule that is composed of two segregated portions, one of which has sufficient affinity for the solvent to bring the entire molecule into solution, and the other portion less affinity for the solvent molecules than the solvent molecules have for each other, can properly be called a surfactant. The list of compounds that fall within that definition would then include at least half of the over six million organic compounds known. How then is one to choose an optimal, or for that matter, even adequate surfactant for dispersing small particles in a given solvent system?

Surface-active compounds that are soluble in organic solvents can be classified according to the following types:

(1) Long-chain polar compounds

(2)   Fluorocarbon compounds

(3)   Silicones

The long-chain hydrocarbons with polar groups do
not lower the surface tension of hydrocarbon liquids.
However, they lower the oil-water interfacial tension
and they are adsorbed on polar surfaces. Typical polar
groups are $-COOH$, $-OH$, $-NH_2$, $-CONH_2$, $-SH$, $-SO_3H$, and
salts of long-chain carboxylic acids and sulfonates.

Short-chain fluorocarbons with polar groups are
frequently sufficiently soluble in hydrocarbon oils to
function as surfactants, lowering surface tension as
well as interfacial tension. Longer-chain fluorocarbons
attached to a hydrocarbon chain of sufficient length are
soluble in hydrocarbon oils, and lower the surface
tension of the oils. They do not lower the interfacial
tension between the hydrocarbon solvent and polar
compounds.

Silicone oils differ broadly in their chemical
structure and surface-active properties. Generally,
they are used as insoluble components of the system and
serve as antifoaming agents. However, those of
sufficiently small molecular weight to be soluble in the
hydrocarbon solvent, and containing only CH_3 groups
attached to silicon in the $(Si-O)n$ skeleton can be
expected to lower the surface tension of the hydrocarbon
solvent. This is due to the lower surface free-energy
of the CH_3 group as compared with the CH_2 group, which
predominates in hydrocarbons.

Surface tension, zeta potential, solubility
parameter, wetting agent demand, adsorption isotherms,
end group analysis, and dispersion stability are all
valuable analytical tools, yet, even if these techniques
were to provide absolute evidence that the surfactant
works with a particular particle in a given solvent-
binder matrix, complete testing of even a fraction of
the potentially useful compounds available would prove
impossible. For that reason, a method of selecting
surfactants useful for dispersing magnetic particles on
a structural basis has been proposed.

The Model Surfactant

In order to prepare stable dispersions of ultrafine
magnetic oxide in organic solvents, a model surfactant

was first proposed by Papell in the early 1960's[1] and later revised by Kaiser[2] and Rosensweig[3] in their attempts to prepare stable magnetic colloids. Although the loading density of the colloids prepared by Kaiser and Rosensweig were about an order of magnitude lower than oxide loading commonly used in dispersions used in magnetic recording media, modifications of the model that they proposed have been useful tools in selecting a surfactant (or more properly stated, an idealized molecular structure) for use in magnetic oxide dispersions.

In general, a molecule of the type

$$R''-R'-R-YH$$

has been observed to yield dispersions with the best magnetic properties, highest degree of dispersion, optimum milling behavior and best performance on tape,[4]

where:

YH is a polar functional group that bonds by means of covalent linkage, chemisorption, adsorption, or ionic interaction to the surface of the magnetic particle. YH is typically COO^-, NH_2, SO_3^-, PO_3^-, M^+Cl^-, or SH, can be mono-, di- or tri-substituted on the terminal carbon atom, such as:

$$R-CH_2-COOH \qquad R-CH(COOH)_2 \qquad R-C(COOH)_3 \; ,$$
$$\text{(a)} \qquad\qquad \text{(b)} \qquad\qquad\qquad \text{(c)}$$

and can be a single or mixed species, such as:

$$R-CH(COOH)_2 \qquad or \qquad R-CH-COOH$$
$$NH_2$$
$$\text{(a)} \qquad\qquad\qquad\qquad \text{(b)}$$

If YH is a cation or anion, it is most effective when the charge of YH is the same magnitude and of opposite sign to the surface charge on the particle being dispersed.

R- is an aliphatic chain (C_4 to C_{20}), aromatic ring or a cyclic aliphatic group. Addition of polar functional groups along the chain (such as primary hydroxyls) can in some cases result in stronger interactions between YH and the particle by acting as additional YH adsorption sites or by increasing the ionization of the YH group. If R is aliphatic, the length of the R chain is useful in changing the magnitude of the charge on YH.

R' is a linking group that causes a change in electron density, separating the polar R-YH portion of the molecule from the nonpolar tail.

R" is a hydrophobic tail that has a similar solubility parameter to the binder/solvent matrix and is of sufficient chain length to separate the magnetic particles, so that no particle/particle magnetic field interference occurs.

Examples of Applications of the Model and Effects of Structural Changes in the Surfactant

It has been found that stable colloidal dispersions of magnetite can be formed using surfactants that adhere to the model. Colloidal dispersions in perfluorinated liquids can be formed by utilizing a fluorocarbon surfactant having the following formula:

$$F\left[\begin{array}{c} CF_3 \\ | \\ C-CF_2-O \\ | \\ F \end{array}\right]_n \begin{array}{c} CF_3 \\ | \\ C-YH \\ | \\ F \end{array}$$

where n is an integer from 3 to 50 preferably 5 to 25; and where R is OOH, OH, $OONH_4$, ONH_2, NH_2, with OOH being preferred.

The stable dispersions of magnetic oxide can be prepared readily. A simple and highly effective technique is to wet grind finely-divided solids in a ball mill in the presence of the surfactant compound dissolved in the carrier liquid, e.g., "Freon" E-3 or the like.

The relative proportions of the surfactant to the suspended solids are not narrowly significant to the preparation of stable colloids. The proportions can be varied widely so long as there is a sufficient concentration of the surfactant component to provide at least a mono-molecular covering of the particles in suspension. The numerical limits in proportions are so broad and dependent on other factors (particle size, density, etc.) as to be almost meaningless, the proportions being, for example, as low as 0.001 parts by weight of surfactant per part of particles for a large, e.g. (1 micron (10,000 A.) low density material to as much as 100 parts by weight per part of very small, e.g. 100 A. particles of a high density material (such as magnetite). A realistic range for forming stable suspensions is 0.01-5.0 parts by weight per part of solids.

A series of colloidal dispersions of magnetite in an inert fluorinated ether carrier were prepared utilizing four different hexafluoropropylene oxide polymer acid surfactants. The solvent utilized was "Freon" E-3 (DuPont) having the following formula:

$$F \left[\begin{array}{c} CF_3 \\ | \\ CF-CF_2-O \end{array} \right]_3 - \begin{array}{c} CF_3 \\ | \\ C-H \\ | \\ F \end{array}$$

while the four surfactants utilized had the following general formula:

$$F \left[\begin{array}{c} CF_3 \\ | \\ CF-CF_2-O \end{array} \right]_n - \begin{array}{c} CF_3 \\ | \\ C-COOH \\ | \\ F \end{array}$$

Surfactant A had a molecular weight of about 1,000, n equal to 5; surfactant B had a molecular weight of 1660, n equal to 9; surfactant C had a molecular weight of 2600, n equal to 15; surfactant D had a molecular weight of 4200, n equal to 24. The magnetite powder utilized was IRN-100 (Charles Pfizer and Co., Inc.). The dispersions were prepared by wet grinding the finely-divided magnetite powder in the presence of surfactant and the inert fluorocarbon carrier using a ball mill half filled with steel balls. The relative proportions of the ingredients used in formulating the dispersions were: Magnetite-1 volume, surfactant-2.5 volumes (4.5 gm.), inert carrier-30 volumes (50 gm.). The ball milling was effective in forming the colloidal dispersion over a milling period of about 1 hour. In each of the four preparations, an effective and permanently stable colloidal dispersion was formed.

In a further investigation of the stability of the four colloidal dispersions, a single drop of each colloid was added to a large excess of test liquid and the compatibility of the colloidal dispersion with the various fluorocarbon media was noted. The fluorocarbon liquids to which the colloidal dispersion was added were Freon E-3, Freon E-5, Freon E-9, Krytox AZ, Krytoz AB, and Krytox AC. These materials have the following formulas:

$$F-\left[CF-CF_2-O\right]_n-CH \quad \begin{matrix}CF_3\\|\\F\end{matrix}$$

$$F-\left[CF-CF_2-O\right]_n-\overset{CF_3}{\underset{F}{C}}-F$$

"Freon E"   "Krytox"

Freon E-3 has a molecular weight of 620, n equal to 3; Freon E-5 has a molecular weight of 950, n equal to 5; Freon E-9 has a molecular weight of 1500, n equal to 9. The Krytox oils AZ, AA, AB and AC have molecular weights, respectively, of 1900, 2500, 3600 and 5000. n in the above formula for the Krytox oils is, respectively, 18, 24, 35, and 49. The results of the tests are summarized in the table, where a (+) indicates that the drop of the colloidal dispersion dispersed readily when added to the indicated liquid, while a (-) indicates that spontaneous flocculation and separation of a solid phase occurred.

Table I

Molecular weight...	Freon			Krytox			
	E-3 620	E-5 950	E-9 1,500	AZ 1,900	AA 2,500	AB 3,600	AC 5,000
A.........	+	+	-	-	-	-	-
B.........	+	+	+	+	-	-	-
C.........	+	+	+	+	+	+	+
D.........	+	+	+	+	+	+	+

The table indicates that the colloidal systems are less stable when the molecular weight of the carrier liquid exceeds the molecular weight of the surfactant by more than about 50%, particularly for the lower molecular weight surfactants. It was observed, however, that, in the cases where flocculation occurred, the addition of either a low molecular weight perfluorinated carrier liquid, such as Freon A-3, or the addition of a higher molecular weight surfactant to the system produced a spontaneous redispersion of the solid phase material.

This instability in high molecular weight solvents is a consequence of not matching the solubility parameter of R" to the solvent matrix.

Colloidal dispersions of magnetic oxide in polyphenyl ether can be formed using a surfactant of the type:

$$\phi - O - \left[(CH_2) \right]_{10} - C \underset{OH}{\overset{O}{\lessgtr}}$$

A series of colloidal dispersions in polyphenyl ether carrier were prepared.

Activated magnetite (FE_3O_4) particles were prepared as follows: 260 grams $FESO_4$, 460 ml 46% $FeCl_3$ and 100 ml water were mixed to dissolve the salts.

800 grams of ice were added to obtain a temperature of -3°C. Fe_3O_4 was precipitated from the solution by the slow addition of a solution of 600 ml concentrated NH_4OH and 400 cc water cooled to +1°C. The Fe_3O_4 was magnetically separated from the salt/ammoniacal solution and was washed with 200-ml portions of hot water, 100-ml portions of acetone and 200-ml portions of xylene and was vacuum dried.

50 grams of the activated magnetite, 15.55 grams of 11, phenoxy undecanoic acid as the surfactant and 525.6 ml of xylene were dispersed for three days in a ball mill. The material was then flocculated with three 100-ml portions of heptane and redispersed with 8 ml of polyphenyl ether (4-ring Nye synthetic oil #433) as a carrier liquid. The product was a stable magnetic colloid with a magnetization of 450 gauss and 1000 cp in viscosity.

A second polyphenyl ether base sample was prepared to study the effect of R" on the stability of the dispersion:

50 grams of activated magnetite, 15.55 grams of undecanoic acid surfactant and 525.6 ml of xylene were dispersed for three days in a ball mill. The material was flocculated wtih heptane as in example #1 and redispersed with 8 ml of 4 ring polyphenyl ether.

The resulting product was a slurry that separated from the carrier liquid within 30 min in a gravity field and spontaneously upon application of a 200 gauss magnetic field.

Although the magnetic colloid model proved useful for studying surfactants in a simple system, it is necessary to understand the interactions of dispersions in a more complex solvent system for preparation of magnetic media.

The Role of the Surfactant in Magnetic Media

The primary role of a surfactant for magnetic media is not unlike that of surfactants used in other particulate dispersions, that is, to disperse particles uniformly in an organic solvent system. Because of the nature of the end product, however, several other factors also must be considered. A typical magnetic tape formulation is:

FIGURE 1

Magnetic oxide	100
Surfactant	3
Polymer binder(s)	25
Solvent	200
Abrasive	1
Carbon black	3
Lubricant	1

Based on this formulation, the conditions in addition to uniform dispersion that the surfactant must satisfy include:

A. Optimization of magnetic properties.

B. Reduction in milling time required to achieve dispersion.

C. Compatibility with the binder system in order to avoid coatings with poor surface quality.

D. Effectiveness at low concentrations in order to avoid softening or plasticization of the binder molecule and to maximize the amount of magnetic material in the coating.

Compatibility with Solvents, Magnetic Particles, and a Wide Range of Polymer Binders

It has been determined that recording media with improved recording properties can be prepared using surfactants that follow the proposed model.

A series of dispersions of Cobalt adsorbed Fe_2O_3 in tetrahydrofuran and a theromplastic urethane were prepared using several surfactants synthesized at Integrated Magnetics for the purpose of this study.

The surfactants prepared were:

A) 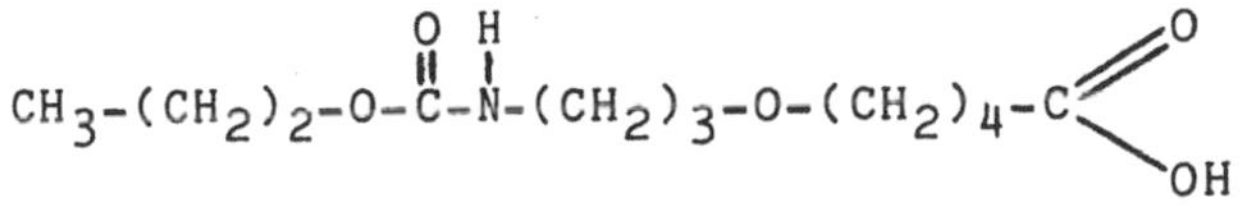 $CH_3-(CH_2)_2-O-\overset{\overset{O}{\|}}{C}-\overset{\overset{H}{|}}{N}-(CH_2)_3-O-(CH_2)_4-C\overset{\diagup O}{\diagdown OH}$

B) $CH_3\left(CH_2-O-\overset{\overset{O}{\|}}{C}-\overset{\overset{H}{|}}{N}-CH_2\right)_2-(CH_2)_3-O-(CH_2)_4-C\overset{\diagup O}{\diagdown OH}$

C) $CH_3\left(CH_2-O-\overset{\overset{H}{|}}{\underset{\underset{O}{\|}}{C}}-N-CH_2\right)_2-(CH_2)_7-C\overset{\diagup O}{\diagdown OH}$

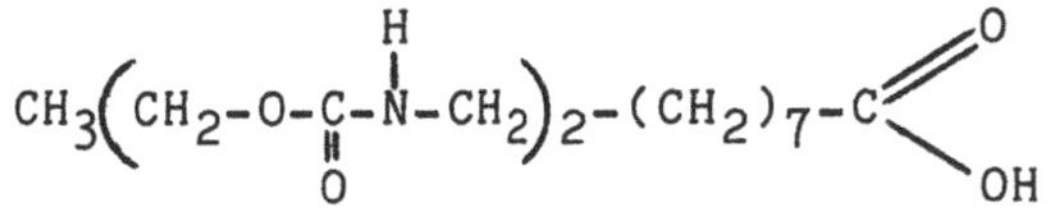

D) 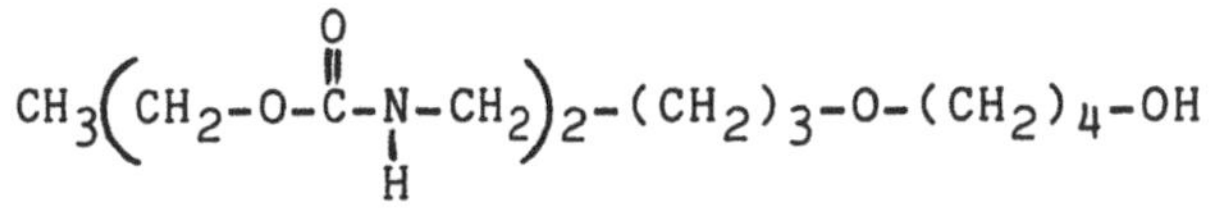 $CH_3\left(CH_2-O-\overset{\overset{O}{\|}}{\underset{\underset{H}{|}}{C}}-N-CH_2\right)_2-(CH_2)_3-O-(CH_2)_4-OH$

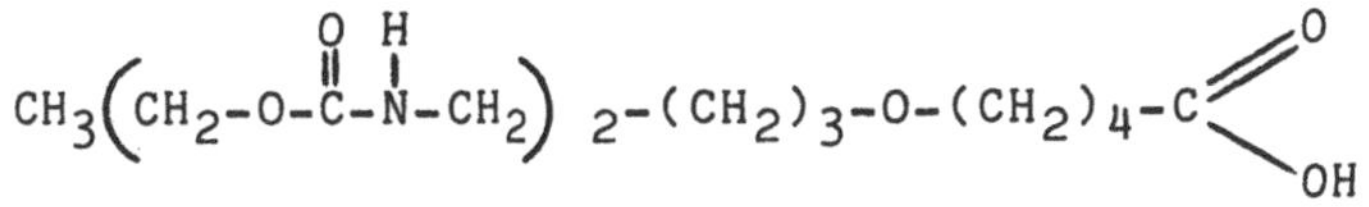

Surfactant A was chosen as a material that fit the model and had an R" tail of moderate solubility in the THF/urethane matrix.

Surfactant B was selected as an ideal surfactant based on the model.

Surfactant C was an analogue to B without an R' linking group.

Surfactant D was an analogue to B with a YH head group with a low affinity for the surface of Cobalt Fe_2O_3.

Dispersions of Cobalt ferrite were prepared in jar mills, according to the formulation given in Figure 1.

Table II

	Surf. A	Surf. B	Surf. C	Surf. D
Coercivity				
1	603	621	560	560
2	615	636	575	570
3	621	636	591	590
4	630	636	599	611
Br/Bm				
1	0.75	0.81	0.42	0.48
2	0.81	0.85	0.45	0.51
3	0.82	0.88	0.51	0.60
4	0.82	0.89	0.56	0.62
Switching Field Distribution				
1	0.48	0.40	0.61	0.58
2	0.46	0.38	0.66	0.56
3	0.46	0.36	0.60	0.50
4	0.46	0.36	0.59	0.50

All samples were prepared in jar mills. The grinding medium used was 4 mm steel balls at a 15:1 media-to-oxide level. All of the oxide, surfactant and 10% of the binder were loaded into the mill in first phase milling. The remainder of the binder was added for a 2nd phase of milling.

For convenience of drying and sample preparation, tetrahydrofuran was selected as the solvent system.

Phase 1 milling was done for a period of 24 hours. The samples were milled for an additional 4 days, after addition of the 2nd phase. Sample recording tapes were prepared every 24 hours, after addition of all of the binder resin.

Table II is a description of the test results of the tapes prepared with the various surfactants.

Conclusions

The selection of a surfactant for magnetic recording media requires a good understanding of the dispersion quality that one wants to achieve, process limitations, and end product usage. Proper selection of the surfactant is necessary to optimize both processing time and tape performance. Selection of a good surfactant can be done based on:

1. Optimization of the polar YH functional head group of the surfactant.

   This can be done experimentally or by choosing a surfactant with a similar magnitude charge and opposite sign to the surface of the particle to be dispersed.

2. Choosing an R" functionality that has a similar solubility parameter to the solvent/binder matrix.

   The solubility parameter of R" can be calculated from Hildebrand units or heats of vaporization as

$$as \quad = \left[\frac{D(\Delta H_v - RT)}{M} \right] \quad 1/2$$

3. Choosing a surfactant molecule that has an R
    chain that minimizes spacing losses between
    particles.

    In general, if R exceeds 20 carbon atoms in
    length, the magnetic signal is decreased
    sufficiently to result in a loss of tape
    output.

References

1. S. Papell, NASA Memo, June 1960.

2. R. Kaiser, U.S. Patent #3784471. Jan. 8, 1974.

3. R. Rosensweig, U.S. Patent #3917538, Nov. 1975.

4. W. Bottonberg and M. S. Chagnon, U.S. Patent
    #4315827, Feb. 1982.

Magnetic Recording
Group Discussion I

Steven A. Snow, Dow Corning Corp., rapporteur

The preparation of dispersions of magnetic oxide particles for use in magnetic recording materials was discussed. Generally, the magnetic particles are dispersed in an organic solvent/binder matrix in preparation for further processing. The role of the surfactant is to act as an effective dispersant for the magnetic oxide particles.

The discussion group addressed both technical and communication issues; the communication issure being one of general concern in applying surfactants to new and emerging technology. It proved impossible to uncouple the technical aspects from the communication aspects, partially due to the pressing need for better communication amongst the variety of workers in the area. These include workers in academic research, in the industrial preparation and distribution of surfactants (suppliers), and the vendors of magnetic recording materials.

On the technical side four issues were covered:
1. There is a need to gain further fundamental insights into the nature of the surfactant-magnetic particle interaction. Development of specific surfactants for this area can only follow improved understanding.

2. A standard for recording tape performance needs to be established. Currently most predictions are made based on subjective
judgment alone.

3. Organometallic-based coupling agents may prove to be useful in
future applications. Especially popular are $Ti(OR)_4$ species.
The coupling agent should chemically bind directly to the magnetic
particle.

4. Polymeric surfactants may also prove to be useful in future
applications.

On the communication side, the major need agreed upon was that
of implementing "outside-in" research, in other words, having
vendors communicate the technical needs of the products to the
surfactant supplier so that surfactants can be "tailored" to
specific customer needs. Ideally, the level of communication
should approach or exceed that of the paint coating industry where
a specific product can be quickly supplied for a specific
application.

Near the end of the discussion, Dr. Chagnon was asked to
summarize where the industry has been, and where he predicts it
might be going. Briefly, the major direction has been toward
higher density storage and will continue to be so in the future.
With higher density storage, risks of static charge destruction
of recorded information are much greater. Through proper surfactant selection, however, the concerns of conductivity co ntrol
may be lessened. In general, greater effort in improving surfactants will be needed in order to allow particulate coating
technology to dominate over the sputtering emission method of
processing magnetic materials.

Magnetic Recording
Group Discussion II

Lesley Nowakowski, American Cyanamid Co., rapporteur

During the early stages of development of tapes and floppy discs, particle density in coatings for magnetic recording media was not critical; the recording density was determined by the limitations of the heads and drives, and research effort was concentrated on those. With improvement in the electronics, higher density discs are now possible (from 0.5 megabytes in 1979 for a 5 1/4 inch floppy to 1.6 megabytes now and an anticipated 10 megabytes in the near future), and traditional coating formulations are no longer sufficient. Previously, the magnetic materials were simply milled in the binder, but now added surfactant is needed to disperse the magnetic particles. The particles are becoming smaller and the role of the surfactant is becoming more critical. Wetting of the particles can be a problem with some of the new materials which are becoming available, because of the high surface area. Particles may also be more difficult to stabilize, due to high sigma values or to shape. Maximum application viscosity for spin coating of rigid discs is 20cps, which limits the use of steric stabilizers. Surfactants are now an indispensible part of the coating formulation, and new surfac-

tants are being sought to accommodate the new magnetic materials
and the need for yet higher recording densities.

The initial role of a surfactant is to deagglommerate particles in the milling process. Afterwards, the function of the
surfactant is to uniformly disperse the particles in the binder/
solvent matrix. To do this it should adsorb well on the surface
of the particle, and also be fully compatible with the binder,
solvent, lubricant, and any other additives. The surfactant
should lower the viscosity (allowing thin coatings and 'write-over'
response characteristics), and keep the particles spaced apart.
A desirable property for the surfactant is that it should behave
in the finished coating 'as though it were not there'. This
could be the case if it also performed in another role: as a
lubricating aid, or as an integral part of the resin system. A
completely different approach would be to create the particles
directly in a non-aqueous medium, in analogous manner to present
aqueous processes. This could result in deagglommerated particles
without the use of surfactant.

The newer magnetic materials can be very different from the
traditional magnetic iron oxide particles. The latter have a
particle size of 1/2 micron, with aspect ratio 10:1 to 3:1, whereas barium hexaferrite (nominally $BaFe_{12}O_{19}$), one of the promising
new materials, is in the form of plate-like particles, approx.
0.05 x .007 microns, with the magnetic axis in the short dimen-
sion. The particles can slide over each other, but tend to stack
like pancakes, which makes wetting difficult. Binders are based
on polyurethanes for tape manufacture, and low molecular weight
epoxy or epoxy-phenolic formulations are used for rigid discs.
These are cured thermally or by amine or anhydride. There is some
experimentation with e-beam curing.

Search for surfactants for use as dispersants for magnetic
particles typically involves screening of hundreds of commercial
samples. Surfactants presently used include certain GAFAC

phosphate ester surfactants. Some of these work, but others do
not. Reasons for good performance of particular surfactants are
not clear. In general, it is known that adsorbing groups are important, in conjunction with the surface chemistry and history
of the surface. Monoesters aid in wetting and diesters give
stability. It would be helpful if more information were available
in the commercial literature on the chemical composition and
structure of surfactants. The surfactant producer, in turn,
would appreciate more information on the use situation in order
better to advise on what surfactants would be appropriate. He
has a wealth of experience on the use of surfactants in different
situations which could be applicable, but he is helpless without
a description of the technical requirements. There is a real
need for better communication which could perhaps be overcome by
secrecy agreements between companies to cooperate on surfactant
development. A drawback is the small market volume. A surfactant
manufacturer would be reluctant to develop a product if the only
application were in the recording media market. An approach being
developed by Pfizer is to sell magnetic particles in a predis-
oersed form. Again, knowledge of the formulation in which the
particles are to be dispersed is important, but a great deal of
proprietary information has been developed, both by the coatings
manufacturers and by the particle producers. Getting a company
such as Nashua to work with a Pfizer may be difficult.

With new particle technology and new surfactant technology,
particulate recording will remain competitive for at least 7-10
years. Its big advantage over thin films (sputtering and plating)
is cheapness (by 2:1). Also, in spite of the promises made by
the thin film technologies in recent years, thin film disc
performance is still not trustworthy and many reconversions to
oxide drives have taken place. Japanese attempts to develop
sputtered tapes have been abandoned because of poor adhesion.
There is now also the long-term emphasis on magneto-optics. In

the future it is expected that each technology will have its place.
Particulate coatings will be used for tapes and floppy discs, and
will remain competitive in the hard disc market. Compact discs
with magneto-optics will be used for write-once applications
such as archival storage, and this may supplant some tape units,
but Winchester disc technology will still be used for data
retrieval.

4

The Use of Surfactants in the Processing of High-Technology Electronic Ceramics

Ellen S. Tormey

Ceramics Process Systems Corp.
Cambridge, Massachusetts

BACKGROUND

In the fabrication of the majority of crystalline ceramics, for electronic or other applications, the starting material is a powder, which is compacted and then sintered to form a dense ceramic. It is now widely accepted that the microstructures which develop during the sintering of ceramics are determined largely by the characteristics of the starting powder and the microstructure of the powder compact (1). To reproducibly manufacture reliable crystalline ceramic components requires control over the characteristics of the starting powder (i.e., particle size, size distribution, purity, morphology) and the forming process such that powder compacts of uniform high green density with the interparticle pore size no larger than a single particle are achievable. For densification of the powder compact with control over the final microstructure, and hence the properties of the sintered ceramic, the variables temperature, pressure and composition must be chosen such that pore removal occurs during sintering and grain boundaries develop between the particles (1).

The fabrication of ceramics often involves processing powders in liquids. The successful processing of ceramic powders in liquids generally requires the achievement of stable particle dispersions. Surfactants are often required to render dispersions stable; their adsorption at the solid/liquid interface promotes the wetting process and contributes to stability via the formation of surface charge and/or adsorbed layers (2). In the processing of electronic ceramics, surfactants can be used to produce powders as well as in the forming operations. In such applications, surfactants are used most often to disperse a powder in a liquid, hence the term dispersant is used by ceramists.

Achievement of a stable dispersion requires the formation of repulsive interparticle forces. In aqueous systems electrostatic repulsion is generally dominant and arises due to the interactions between the electric double layers surrounding the dispersed particles. In nonpolar organic media (e.g., hydrocarbons) stability arises due to repulsion between interacting molecules adsorbed onto the particle surfaces and is generally referred to as steric stabilization. As a general rule, in the latter system, the most effective dispersants have strongly adsorbed functional groups and strongly solvated chains which extend into the solvent. Systems which are stabilized by a combination of mechanisms (i.e., charge and steric) tend to be the most stable (2).

In ceramics powder processing, steric stabilization offers several advantages over electrostatic stabilization (3). Sterically stabilized systems are less sensitive than electrostatically stabilized ones to trace additives or impurities. Electrostatically stabilized systems are very sensitive to electrolyte concentration; too much electrolyte can result in flocculation due to collapse of the electric double layers surrounding the dispersed particles. Steric stabilization can be effective in both aqueous and nonaqueous media, whereas the electrostatic mechanism is generally only effective in water or polar organic solvents. Most importantly, steric stabilization is effective in dispersions containing high volume fractions of solids, typically used to process ceramics, whereas electrostatic stabilization is generally only effective in dilute systems.

INTRODUCTION

The ceramic materials used for electronic applications are generally oxides, but also include carbides and nitrides. The compositions, properties and applications of some of the more common electronic ceramic materials are given in Table I. The surfaces of most oxide and even non-oxide powders have similar chemistries in that they are normally hydroxylated. However, the liquids used in processing such powders can vary greatly and include water, polar and nonpolar organic solvents or mixtures thereof. Hence the type of surfactants used to disperse ceramic powders can vary widely depending on the specific powder/liquid system. The general requirements for dispersants used in the processing of electronic ceramics are as follows: (1) they must be compatible with the other components used in the processing, such as binders and plasticizers; (2) they must burn out cleanly and easily during firing leaving no residual carbon or other

TABLE I. Electronic Ceramic Materials

MATERIAL	PROPERTY	APPLICATION
Al_2O_3, BeO, SiC, AlN, Si_3N_4, BN, $2MgO \cdot 2Al_2O_3 \cdot 5SiO_2$, Borosilicates	Insulation	Substrates, IC Packages, Insulators
$BaTiO_3$, $SrTiO_3$	Dielectric	Capacitors
TiO_2, ZrO_2, SiC, ZnO	Semiconductive	Thermistors, Sensors, Varisitors
ZrO_2, β-Al_2O_3	Ion Conductive	Oxygen Sensors, Ion Sensors, Fuel Cells
$Pb(Zr,Ti)O_3$, $BaTiO_3$, ZrO_2	Piezoelectric	Oscillators, Ultrasonic Elements
γ-Fe_2O_3	Magnetic	Magnetic Disks

contaminants which could adversely affect electrical properties and/or interfere with the sintering process; and (3) they must be inexpensive.

In what follows are some specific examples of the use of surfactants as dispersants in the processing of electronic ceramics. The discussion includes examples of the use of surfactants in powder production as well as in forming processes.

POWDER PRODUCTION

The first step towards obtaining a green powder compact of uniform high density is to produce or synthesize an "ideal" powder, the characteristics of which are as follows: small size (<1μm), narrow size distribution (monosized), nonagglomerated, equiaxed, and phase and composition pure (1). Narrow sized powders can be obtained by either of two basic techniques; via size classification of commercially available finely ground wide size distribution powders or precipitation of fine particles from a solution or gas phase (4).

Sedimentation of particles dispersed in a liquid has proven to be an effective technique for the separation of wide size distribution powders into narrow sized fractions (4). The technique is based on the principle that the sedimentation rate of the dispersed particles is proportional to the square of the particle radius, according to Stokes law. In order for such a sedimentation technique to be effective a well dispersed system is required such that each particle is separated from every other particle, insuring that the particles settle individually rather than as agglomerates. Dispersants used in the size classification of powders must be compatible with any additives (such as solvents, binders,etc.) used in subsequent processing steps; otherwise, they must be easily removed from the powder surfaces prior to further processing. For safety and environmental reasons, water or polar organic solvents such as alcohols are the most attractive media for classifying powders.

Ceramic powders may be size classified in water without a surfactant, provided the water has a very low electrolyte concentration (<0.01 molar) and is at a pH several units above or below the isoelectric point of the solid particles (4). For example, a commercially available Al_2O_3 powder has been successfully classified into various submicron sized fractions using a sedimentation technique, when dispersed in water adjusted to a pH of 3-4 (5).

Although the use of a dispersant is not necessary to promote stable dispersions of oxide powders such as alumina in water, the use of a dispersant which contributes to surface charge and also provides a steric barrier to flocculation is likely to enhance dispersion and thus make the classification process more effective. One such class of dispersants are the polyelectrolytes; in particular, ammonium polyacrylates have been used widely in the ceramics industry for dispersing oxides such as alumina in water. Some water soluble polymeric type dispersants, that can also act as binders in subsequent processing steps have been found to be effective for dispersing oxides in water. For example, polyvinylpyrrolidone has been shown to be a good dispersant for alumina (6) and has also been used as a binder in aqueous casting systems for alumina (7). In water and polar organic solvents such as isopropanol, low molecular weight organic acids such as para-hydroxybenzoic and para-aminobenzoic acids have been found to be effective dispersants for oxide powders (8,9).

The controlled hydrolysis of metal alkoxides in solution has emerged in recent years as a technique for synthesizing submicron, spherical monosized oxide powders such as SiO_2, TiO_2, ZrO_2 and ZnO (10-12). Bagley (13) is now investigating the hydrolysis of alkoxides in emulsions for forming narrow sized oxide powders containing multiple cations. The reason for using an emulsion is that if an alkoxide solution can be stabilized as dispersed droplets in a second immiscible liquid using a surfactant, then each droplet can act as a microreactor and the overall particle composition, shape, size, and size distribution are controlled by the emulsion.

Bagley, in her initial studies, formed spherical SiO_2 particles from emulsions. Stable water in heptane (W/O) emulsions were formed, to which an alkoxide (tetraethylorthosilicate) was added; HCl gas bubbled through the emulsion catalyzed the alkoxide hydrolysis reaction, resulting in the formation of SiO_2 particles and alcohol as a byproduct. Bagley found that the more stable the emulsion, the higher were the powder yields, and that surfactants which formed stable W/O emulsions in the presence of alcohol, without reacting with the alkoxide, were required. Nonionic surfactants such as sorbitan trioleates and oleic diethanolamide were assessed as emulsion stabilizers; the oleic diethanolamide resulted in more stable emulsions in the presence of alcohol and also better powder yields than did the sorbitan trioleates (13,14).

FORMING PROCESSES

Once a narrow sized powder is obtained, one needs to compact the powder particles to form a green microstructure of uniform high density. Ceramics processing techniques such as sedimentation, slip casting, tape casting and electrophoretic deposition can result in greenware of uniform high density when narrow sized particles are packed from fully dispersed (stable) suspensions.

One can appreciate the need for a well dispersed system in ceramics forming operations in terms of the sedimentation behavior of particle dispersions. The relative stability of a dispersion affects the settling and packing behavior of the particles (2). In general, the particles of a flocculated (unstable) dispersion settle rapidly in bulk to form a loosely packed sediment of large specific volume, which is easily redispersed. In contrast, the particles of a deflocculated (stable) dispersion settle slowly as individual particles and the final sediment is densely packed and not easily redispersed. Hence, the greater the degree to which the particles are dispersed in the liquid, the denser and more uniform will be the particle packing in the formed piece.

The tape casting process is commonly used to form thin flat ceramic sheets and is used to manufacture electronic ceramics such as substrates and multilayer capacitors and integrated circuit packages. In the tape casting process, dispersants are used in conjunction with solvents, binders and plasticizers to form a slurry which is leveled beneath a doctor blade to form a thin ceramic sheet. A typical tape casting formulation for alumina substrates is that developed by Shanefield and Mistler (15) and which is given in Table II. The dispersant in this particular system is Menhaden fish oil, a naturally occurring oil composed

TABLE II.   Tape Casting System (15)

MATERIAL	FUNCTION	PARTS BY WEIGHT
Alumina Powder	Substrate Material	100.00
Magnesium Oxide	Grain Growth Inhibitor	0.25
Menhaden Fish Oil	Dispersant	1.70
Trichloroethylene	Solvent	39.00
Ethyl Alcohol	Solvent	15.00
Polyvinyl Butyral	Binder	4.00
Polyethylene Glycol	Plasticizer	4.30
Octyl Phthalate	Plasticizer	3.60

primarily of glyceryl esters of saturated and unsaturated fatty acids of chain length varying from 14 to 22 carbons. Menhaden oil has been found to be an effective dispersant for many oxide powders dispersed in nonpolar organic solvents (16,17). Studies have shown that fish oil is an effective dispersant due to the presence of carboxylic acid groups along the triglyceride chains, formed as a result of oxidation of the oil during its processing (18,19). The carboxylic acid groups anchor strongly to the hydroxylated particle surfaces, while the long flexible chains dissolve freely in the solvent providing a good steric barrier to flocculation. For a dispersant to be effective in a tape casting system it must compete with the solvents, binders and plasticizers for the particle surfaces and at the same time be compatible with these other system components.

One can envision a number of improvements to this tape casting system in terms of the dispersant (20). Fish oil, because it is a naturally occurring substance, has a variable composition and purity. This points to the need for a synthetic analog of fish oil which is similar to the oxidized oil in structure but pure and reproducible. Actually, a number of commercially available synthetic substitutes for fish oil have been found (17), but probably none have the same chemical structure as oxidized fish oil. It may also be desirable to increase the chain length of the molecule to the point where the dispersant also acts as a binder.

The dispersants typically used in ceramics processing bond to the particle surfaces via hydrogen bonding or weak chemical bonding. Dispersants which can chemically react (couple) with the particle surfaces to form stronger bonds offer distinct advantages in powder processing (3). Formation of a strong surface chemical bond would ensure that the dispersant remains on the particle surfaces during subsequent processing steps, resulting in a system which is less sensitive to slight compositional or processing condition variations. Coupling agent type dispersants would be especially advantageous in tape casting systems, since they are multicomponent and competitive adsorption is likely to be operative.

Parish (3) and Lalanandham (21) have investigated the use of low molecular weight organotitanates as dispersants for ceramic powders (BaTiO$_3$, SrTiO$_3$, Al$_2$O$_3$) in nonpolar organic solvents such as hexane and toluene. When an organotitanate, such as diisopropoxy titanium dioleate, is combined in a nonpolar organic solvent with an oxide powder, the organotitanate isopropoxy groups react with the particle surface hydroxyls forming isopropanol and a doubly surface bonded titanium dioleate molecule according to the following reaction (3):

$$(oleate)_2\text{-Ti-}(isoproxide)_2 \ + \ (HO)_2\text{-surface} \ \rightleftharpoons$$

$$2iPrOH + (oleate)_2\text{-Ti-}(surface)_2$$

The long chain oleate ligands of the covalently attached molecule extend into the solution phase, providing a steric barrier to flocculation. The adsorption of organotitanate dispersants has resulted in stable dispersions of oxide powders in nonpolar organic solvents when a monolayer or more of the dispersant was present on the powder surfaces (3,21). Lalanandham (21) has found that an organotitanate was a better dispersant for Al$_2$O$_3$ in hexane or toluene than was the Menhaden fish oil.

There are a number of coupling agents which are commercially available for use as dispersants, e.g., the organotitanates and organosilanes. If not commercially available, conceivably, one could synthesize such a dispersant by combining the appropriate alkoxide and organic ligand, thereby tailor making the dispersant for a given system. When using coupling agents as dispersants for electronic ceramics, the metallic portion of the molecule must not be detrimental to the electrical properties of the ceramic or interfere with the sintering process, since this portion is not removed from the body during sintering. In the case of a metal-oxygen linked organic chain, the dispersant will decompose during sintering to a metal oxide residue.

In theory, coupling agents can serve not only as dispersants in ceramics powder processing, but also as dopants and/or binders. Dopants (generally secondary metallic oxides) are often added to ceramic powders to aid in the sintering process. For example, MgO is a well known sintering aid for Al$_2$O$_3$; likewise, Y$_2$O$_3$ is commonly used to enhance the densification of AlN (22). The addition of such a dopant in the form of a coupling agent would ensure that it is homogeneously distributed in the green body and also the sintered ceramic, since it bonds to the particle surfaces. For use as binders, coupling agents with polymerizable ligands can be synthesized (21). Such a molecule could first act as a dispersant and then be converted to a binder via an in situ polymerization step.

SUMMARY

To reproducibly manufacture reliable electronic ceramics requires the use a narrow sized starting powder and the formation of a uniformly and densely packed green compact. The key to obtaining a green compact consisting of uniformly and densely packed narrow sized particles is too adequately disperse the ceramic particles in a liquid phase during processing. Surfactants are generally used as dispersants in electronic ceramics processing. They are of use in powder synthesis, in the size classification of commercially available ceramic powders, and in forming operations such as tape casting.

The future will bring the increased use of surfactants in electronic ceramics processing, not only as conventional type dispersants, but as coupling agents, dopants, binders and wetting agents.

ACKNOWLEDGEMENTS

The author thanks Mark Parish and Raysha Picerno for their assistance in the preparation of this paper.

REFERENCES

1. H.K. Bowen, Matls. Sci. Eng., **44** (1980) 1.
2. G.D. Parfitt, Ed., The Dispersion of Powders in Liquids, 2nd Ed., John Wiley & Sons, New York, 1973.
3. M.V. Parish, Ph.D. Thesis, MIT, Cambridge, 1985.
4. R.L. Pober, E.A. Barringer and H.K. Bowen, ISHM Technical Monograph Series 6984-004, International Society For Hybrid Microelectronics, Silver Spring, MD, 1984.
5. R.L. Pober, R. Hay and M.L. Harris, to be published.
6. CPRL Report #Q5, April-June 1985, Ceramics Processing Research Lab, MIT, Cambridge.
7. M. Kemr and H. Mizuhara, U.S. Patent 4,329,271, 1982.
8. CPRL Report #R8, Jan.-March 1986, Ceramics Processing Research Lab, MIT, Cambridge.
9. M.V. Parish, M.S. Thesis, MIT, Cambridge, 1982.
10. W. Stober et al., J. Coll. Int. Sci., **26** (1968) 62.
11. E.A. Barringer and H.K.Bowen, J. Amer. Cer. Soc., **65** (1982) C199.
12. R.H. Heistand II et al., Proc. Int. Ultrastructure Conf., University of Florida, Gainsville, Feb. 1985.

13. A. Bagley, CPRL Report #54, Ceramics Processing Research Lab,
    MIT, Cambridge, 1985.
14. CPRL Report #R7, Oct.-Dec. 1985, Ceramics Processing Research
    Lab, MIT, Cambridge.
15. D.J. Shanefield and R.E. Mistler, Amer. Cer. Soc. Bull., **53**
    (1974) 416.
16. E.S. Tormey et al., in <u>Surfaces and Interfaces in Ceramic and
    Ceramic-Metal Systems</u>, J. Pask and A. Evans, Eds., Plenum,
    1981, p.121.
17. K. Mikeska and W.R. Cannon, Advances in Ceramics, Vol. 9, The
    American Ceramic Society, Columbus, 1984, p.164.
18. E.S. Tormey, Ph.D. Thesis, MIT, Cambridge, 1982.
19. P.D. Calvert, E.S. Tormey and R.L. Pober, Amer. Cer. Soc.
    Bull., **65** (1986) 669.
20. E.S. Tormey et al., Advances in Ceramics, Vol. 9, The American
    Ceramic Society, Columbus, 1984, p.140.
21. CPRL Report #Q3, July-Dec. 1984, Ceramics Processing Research
    Lab, MIT, Cambridge.
22. K. Komeya, Amer. Cer. Soc. Bull., **63** (1984) 1158.

High-Technology Electronic Ceramics
Group Discussion I

J. A. Wingrave, E. I. du Pont de Nemours, rapporteur

Preparation Method: High-technology ceramics are prepared by kiln firing a green tape. The green tape is made from a mixture of a polymer solution and an inorganic powder. The green tape results when the solvent is removed from the mix. The polymer is thermally removed in the firing process along the grain boundaries, leaving the ceramic.

Dispersion Methods: The polymer dispersants stabilize the mix by steric or electrostatic forces. While steric is preferred the latter method can be used so long as undesirable cations are not left behind in the finished ceramic. Electrostatic stabilization is also more sensitive to pH changes, but do not preclude this method so long as pH can be properly controlled. While dispersants are in widespread use, more basic research into how dispersants behave in ceramic pastes and green tapes is needed. The high-technology nature of this field has impeded research since companies producing high-technology ceramics are hesitant to share research and technology with academia or other companies.

Requirements for a good ceramic: The specific requirements depend
on the particular application. Examples of the breadth of requirements for high-tech ceramics would be:

Density,

Strength,

Electrical properties,

Surface smoothness,

Crystallinity,

Agglomeration,

Fiber incorporation,

Dopant incorporation and,

Particle diameter and distribution.

Problems in Ceramics Industry: Ceramics production in high-tech
applications involves many different technologies; i.e., ceramics,
colloid chemistry, combustion chemistry, polymer chemistry,
electrical engineering, etc. In addition, many grades and formulations are required for the plethora of applications. As a result
research into ceramic technology is difficult to justify
economically.

Under current technology, many ceramic powders are ground in
order to achieve proper particle size. However, the high purity
needed in high-tech ceramics required very special grinding
procedures. Areas in which further research are needed are:

grinding mechanisms in ultra-fine powders,

behavior and mechanisms of grinding aids,

effects and sources of contaminants,

analytical tests for contaminants at low concentration levels,

forces between particles during grinding, as well as in

dispersion and sintering,

interaction when a powder contains two or more minerals, from

grinding through processing and to end use application,

surface vs bulk properties of powders.

As a result of many of these problems, most U.S. high-technology ceramic consumers currently buy their ceramics from
overseas.

Summary

   High-technology ceramics are a large and growing industry typified by a huge number of relatively small volume products. This market character has allowed high-tech ceramic applications to outstrip the fundamental understanding and research resources of these applications. If the high-tech ceramics industry is to rapidly develop the full potential possible for these materials in the U.S., basic, multidisciplinary studies must be encouraged in both academia and industry. The proprietary nature of these tech-nologies will make it difficult to achieve these ends in the U.S. alone.

High-Technology Electronic Ceramics
Group Discussion II

Carolyn A. Ertell, Stauffer Chemical Co., rapporteur

Three important general points were emphasized throughout the
discussion.

1. Technology overlap: The conference attendees, and thus our
Discussion Group members, were drawn from many different areas of
expertise, academic or industrial. Yet, we were struck by the
overlap of our diverse areas of training and strength and the
ease with which we could bring to bear our ostensibly separate
backgrounds to solve problems in a different high-technology
application area. For example, the technology of minerals
processing, and dispersion, flocculation, and stabilization theory
and know-how could be applied to high-technology ceramics
processing.

2. Communication: A conference such as this provides a special
opportunity to meet with other scientists who have similar interests but are in different specialties. Even though we may have
common interests, we probably don't read the same journals, belong
to the same Societies, attend the same meetings, or even talk with
each other. Interaction such as this Conference is beneficial.

3. Research Needs: While some fundamental principles may be
known and some specific, demonstrated applications may be well

known, there appears to be a need for further R & D in that in-between area which encompasses the bench-scale and scale-up application of surfactant technology areas such as electronic ceramics. This can be a particularly difficult area for which to gather information. Much surfactant applications information is obtained from suppliers' technical specialists. However, their available information is generally geared to the high volume use of surfactants. For example, surfactant suppliers provide excellent information on fundamental surfactant studies or information on detergency studies useful to "Soapers," but often do not have information on the use of surfactants in traditionally low-volume areas such as electronic ceramics.

Some specific needs for the electronic ceramics area are reducing the binder level, employing tight raw material and process control, and obtaining surfactants of definite composition and purity with no lot-to-lot variation.

5
Surfactants in Advanced Battery Technology

Patrick G. Grimes

Exxon Research and Engineering Co.
Annandale, New Jersey

I. INTRODUCTION

Surfactants play an important and critical role in advanced battery
technology. Surfactant type compounds are used to improve plating
of metals onto electrodes, to complex reactants and to change
surface properties of battery components.

Surfactant use in advanced battery systems can be discussed by
considering zinc batteries and, in particular, zinc/bromine battery
systems. Several companies and agencies have investigated
zinc/bromine systems (1-18).

The zinc/bromine battery systems can be illustrated in the
circulating electrolyte system under development by Exxon Research
and Engineering.

II. ZINC/BROMINE BATTERIES

Zinc/Bromine Battery development efforts at Exxon have recently
concentrated on scaling-up the technology to 20- and 30-kWh systems

in order to better demonstrate the technical feasibility of the design approach. In the beginning 20-kWh, 120-volt batteries were designed for energy storage and stationary storage evaluation. This design approach has recently been modified to build a 30-kWh, 200-volt battery suitable for testing with an EV (electric vehicle). The development of bulk energy storage and electric vehicle markets has been severely limited by the lack of better energy storage batteries. Lead-acid batteries presently dominate these markets. However, lead acid batteries suffer limitations in the areas of weight, cost, and maintenance.

Zinc/bromine batteries offer significant improvements over present lead-acid battery technology. Furthermore, results from recent progress on the zinc/bromine system have been positive, and the system is now considered to be an outstanding advanced battery system. Commercially available materials can be used to make the electrolyte components. The battery structures are almost entirely plastic, including plastic electrodes. The extensive use of plastic allows for light weight, low cost, and efficient mass production of battery components.

The performance of present electrodes and electrolytes in parametric batteries allows projections to 60 Wh/kg and 130 W/kg in a "typical" advanced design. Furthermore, there is great design flexibility in this approach, so that the specific energy or power emphasis could be extended to either 80-Wh/kg or 205-W/kg. By varying the bipolar stacking, battery packages have been designed for 36- to 240-volts. Circulating zinc/bromine battery designs could realistically be scaled between 10-kWh and 100-kWh to meet specific applications scenarios.

A. BATTERY OPERATION

The overall operation of the zinc/bromine battery is based on a circulating electrolyte, as shown in Fig. 1. Electrolyte circulation is particularly useful for feeding reactants, removing products, assisting thermal management, and homogenizing the

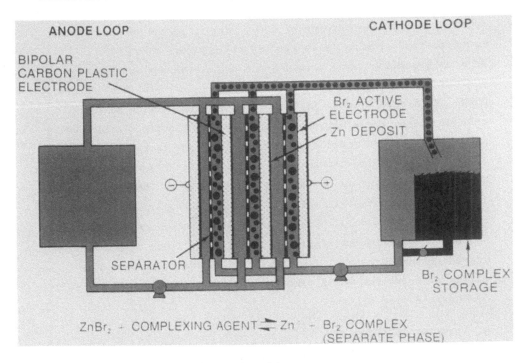

FIG. 1   *Zinc/bromine battery schematic.*

electrolyte. Although electrolyte circulation increases design
complexity, it frequently results in improved or enhanced
performance and higher specific energies. A particular advantage
in the zinc/bromine battery is the improved uniformity of the zinc
plating during charging.

The principal component of the battery is the electrochemical
module, where the actual electrochemistry takes place. The second
component is the electrolyte, which is an aqueous solution of zinc
bromide, supporting salts, bromine complexing agents and
additives. The electrolytes are circulating in two streams through
the electrochemical module. The third component is the system of
pumps and reservoirs which circulates and stores the electro-

lytes. During the charging process, zinc is plated at the negative electrode and bromine is evolved at the positive electrode. Bromine associates with bromide ions and quaternary ammonium ion surfactant type complexing agents to form a low solubility second phase, indicated by dots in Fig. 1. This bromine-rich phase is carried out of the electrochemical module by the circulation and is separated from the aqueous phase by gravity and stored in the catholyte reservoir. Long-term charge retention is excellent because the bromine is stored away from the zinc. During discharge, the catholyte valve is opened and the bromine complex is fed back to the module. Zinc and bromine react electrochemically to reform the original zinc/bromide solution, liberating the energy stored during charging. The separator is used to prevent direct mixing of the circulating electrolyte loops, thereby reducing self-discharge during cycling.

B. BATTERY CONSTRUCTION

Stack design utilizes the 1200 cm^2 two-piece unit cell. This includes co-extruded bipolar electrodes, which are shown in Fig. 2. In these electrodes, a central conductive plastic is bordered during the extrusion process by two non-conductive plastic strips. Non-conductive plastic strips are needed at the extremities of this electrode because of shunt current considerations. The injection-molded separator flow frame used with the larger co-extruded bipolar electrodes is also shown in Fig. 2. In this separator, an insert of microporous plastic with raised posts is surrounded by non-conductive plastic which includes various flow distribution channels. These channels direct and distribute the two electrolytes over the faces of its adjacent bipolar electrodes. The flow channels have been designed to utilize high-conductivity electrolytes and tunnel shunt current protection, as discussed in detail elsewhere (19,20). An exploded view of a stack of cells is shown in Fig. 3. In a typical design, a common central manifold is located between the positive end of the two stacks. The extreme ends of the two stacks can, therefore,

FIG. 2 *Electrode and separator components.*

have a negative polarity and the stacks would be operated in
parallel. The external negative end blocks, as shown in Fig. 2,
are molded from a reinforced polyolefin. The stacks, the central
feed block and the two end blocks are compressed together by
metallic tie rods. However, the actual sealing of the stacks is
done with a sealant which bonds the parts together.

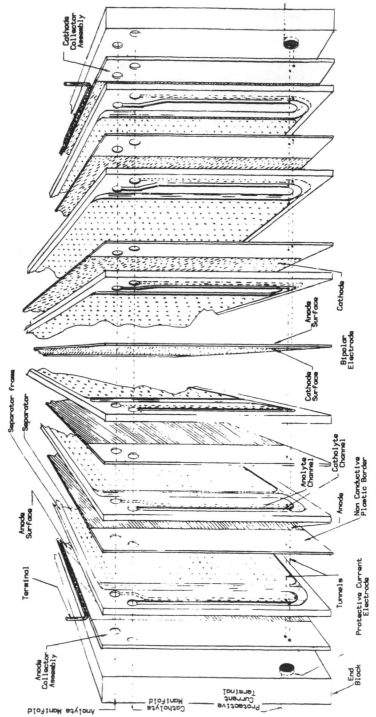

FIG. 3 Exploded view of battery stack.

Labels in figure:
Cathode Collector Assembly
Anode Surface
Cathode
Bipolar Electrode
Cathode Surface
Separator Frame
Separator
Catholyte Channel
Anolyte Channel
Anode
Non Conductive Plastic Border
Anode Surface
Terminal
Protective Current Electrode
Tunnels
Anode Collector Assembly
End Block
Protective Current Terminal
Catholyte Manifold
Anolyte Manifold

II.   SURFACTANT USE IN ZINC DEPOSITION

The achievable capacity of zinc batteries is limited by the thickness and morpohology of the zinc deposit formed during the
charging. There is a great tendency to grow fern or Christmas-tree-
like dendrites on the zinc metal surface during the plating process
of charging the battery. Severe dendrite formation can produce
internal electrical shorts in the battery, thus lowering
efficiency.

The morphology of the zinc electrode surface can be improved
by the flow of the zinc ion containing electrolyte over the
electrode surface (21). Higher flow rates are needed to reduce or
eliminate dendrite formation at higher deposition current
dendrites. The work of Naybour on the morphology of zinc as a
function of Reynolds number, flow velocity and current density is
shown in Fig. 4. While the flowing electrolyte helps in producing
smooth zinc plate, dendrite inhibiting additives are useful in
producing high thickness zinc loading or static situations.

Zinc plating from acidic chloride electrolytes has been
extensively studied by MacKinnon and Brannen (22). They studied
the effects of organic additive on the character of the zinc plate
relative to fine grain, smoothness and compactness of the deposited
zinc sheet for electrowinning of zinc metal. This work continues
in zinc halogen batteries as seen later.

It was found that
1. Positively charged additives were effective and they
    migrated to the cathode and were absorbed.
2. The quantity of additive ion in the deposit was small and
    difficult to confirm analytically.
3. Columnar type deposit growth is eliminated by the additive
    presence.

Screening studies were made to evaluate the effect of organic
additives on the 24-hour zinc deposits from acidic zinc chloride
solutions. The additives studied were:

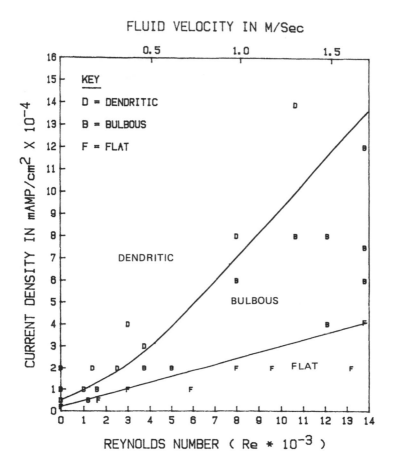

FIG. 4   *Effect of flow rate, Reynolds number, and current density on zinc morphology.*

TEACl; tetraethylammonium chloride

TPrACl; Tetrapropylammonium chloride

TBACl; Tetrabutylammonium chloride

TPACl; Tetrapentylammonium chloride

Pearl Glue; animal glue molecular weight, 50000

Percol 140; high molecular weight cationic polyacryamide

(Allied Chemical)

Sepran NP10; high molecular weight polyacryamide (Dow Chemical)
The electrolyte was maintained at 0.2M $ZnCl_2$ and 0.12M HCl during
the deposition. The current efficiency, energy requirements and
zinc orientation for the various additives are shown in Table 1.
The tetrabutylammonium chloride was most effective at 15 mg/liter
of TBACl, the deposit was refined dense zinc platlets aligned
perpendicularly to the electrode surface. The glue and the
polyacryamide complex compounds which are effective with normal
acidic zinc sulfate electrowinning solutions were less effective in
the chloride system.

The homologous series of tetraalkylammonium chloride was
investigated in view of the good results of TBACl. The data for
the series is shown in Table 2. The concentration of additives was
in the range of 15-60 mg/ liter. The ethyl form was the least
effective. While the propyl and pentyl forms were effective giving
good deposits, the efficiencies were lower. The tetrabutylammonium
chloride additive produced the best deposits with the greatest
efficiency.

The deposit orientation became more platlet in form and
perpendicular to the surface as the alkyl chain length increased,

TABLE 1 *Current efficiency, energy requirement and orientation data for
zinc deposits as a function of various additives*

Additive (mg/ℓ)	CE (%)	ER (kWh kg^{-1})	Orientation*
Addition-free	91.0	3.8	112, 114
TBACl (7.5)	91.0	3.9	110
TBACl (15)	99.4	3.7	101, 110
Pearl glue (30)	91.1	3.6	100, 110
Pearl glue (60)	96.9	3.1	110, 100
Percol 140 (6)	93.0	3.4	112, 110
Separan NP10 (6)	94.3	3.3	110

Relative to ASTM standard for zinc powder.

*Table 2 Current efficiency, energy requirement and orientation data for
zinc deposits as a function of various tetraalkylammonium chlorides*

Additive (mg/ℓ)	CE (%)	ER (kWh kg^{-1})	Orientation*
Addition-free	91.0	3.8	112, 114
TEACl (15)	91.2	4.0	114, 103
TEACl (30)	88.9	4.1	114, 103
TEACl (30)	92.4	4.0	112, 114
TEACl (60)	93.2	4.7	114, 112
TPrCl (15)	93.0	4.1	110
TPrCl (30)	95.1	4.2	110
TPrCl (45)	97.2	3.7	103
TPrCl (60)	97.9	4.0	114, 112
TBACl (7.5)	91.0	4.0	110
TBACl (15)	99.4	3.7	101
TBACl (15)	96.3	3.2	101, 110
TBACl (15)	94.6	4.0	-
TBACl (15)	96.7	3.7	110
TPACl (10)	92.7	4.2	110
TPACl (15)	94.8	3.9	110
TPACl (30)	95.3	3.8	110
TPACl (45)	95.8	4.3	110

Relative to ASTM standard for zinc powder.

indicating greater polarization. Dingle and Damjanovic (23) found
that the deposits were less dendritic and smoother from an alkaline
zincate solution when the chain length increased from tetramethyl
to tetrapentyl in tetraalkyl ammonium bromides.

MacKennon and Brannen state that organic additives are
essential to smooth deposits of zinc from acidic chloride
solutions.

Several proprietary organic additives have been used in the
zinc/chlorine battery system by Energy Development Associates
(24). These are of the surfactant family. Unfortunately these

organic additives are oxidized by the battery environment (acidic solutions saturated with chlorine). These oxidization reactions alter the electrolyte pH and can lead to periodic electrolyte replacement or purification (25). Rigorous electrolyte purity requirements and inorganic additives have to be used in the zinc chlorine battery system (25).

In the investigation of the zinc/bromine battery system, Gould, Inc. looked at a series of quaternary ammonium compounds which have surface active properties and are stable in the presence of bromine (26). They were chosen to be good dendrite inhibitors and also be bromine complexing agents. The particular specific quaternary ammonium compounds have not been identified by Gould.

Good results with quaternary ammonium compounds ZDS-2 and ZDS-3 are shown in Table 3. The concentrations of additive in the electrolyte ranged from 0.001M to 0.05M. The carbon electrodes in the studies were plated with 200 mAh/cm^2 (0.013 in.) and 300 mAh/cm^2 (0.020 in.) loadings of zinc in flowing electrolyte cells at 40 mA/cm^2 rate. The electrolyte was 2M ZnBr$_2$ and 4M KCl.

Gould (27) and General Electric (28) have studied fluorocarbon surfactants as zinc dendrite inhibitors. The Gould study was done with the addition of 100 ppm of Zonyl® FSN (Dupont) to the zinc bromide electrolyte. The effect of loading (200 and 300 mAh/cm^2) and titanium and carbon substrates were investigated. The results are shown in Table 4.

Meidensha Electric Company (Japan) has found quaternary ammonium compounds with one alkyl group of 10-20 carbons to be very effective dendrite suppressors. The quaternary ammonium compounds in conjunction with small quantities of tin and lead salts show a synergism (29).

The quaternary ammonium compounds, N-methyl, N-ethylmorpholinium bromide and N-methyl, N-ethylpyrrolidinium bromide used in the Exxon zinc/ bromine battery system at the 1 molar total level act as dendrite suppressors and as bromine complexing agents. These complexers have been used in systems that

Table 3 Zinc deposition studies: quat effect quality ratings

	No Additive	ZDS-2 .001M	ZDS-2 0.1M	ZDS-2 .05M	ZDS-3 .001M	ZDS-3 .01M	ZDS-3 .05M
			Loading	= 200 mAh/cm^2			
Ratings	56	66	100	100	40	46	87
	45						
	53	61	100	99			84
	48						
	47	60	99	102	39	43	74
	40						
	43	71	101	101			82
	47						
Average	47	65	100	100	40	44	82

	No Additive	ZDS-2 .001M	ZDS-2 0.1M	ZDS-2 .05M	ZDS-3 .001M	ZDS-3 .01M	ZDS-3 .05M
			Loading	= 300 mAh/cm^2			
Ratings	0	23	98	85	9	0	60
	0	0	88	86			47
	0	20	96	103	8	1	55
	0	18	99	92			60
Average	0	15	95	92	8	1	56

have cycled over 1500 cycles without significant change in the electrolyte composition.

Many organic additives have been proposed and investigated to improve zinc electrochemical plating. The quaternary ammonium compounds appear to be the most effective. Further studies will lead to more stable and optimum formulations.

In studies of alkaline silver/zinc cell, the incorporation of surfactants such as tridecyloxypoly(etheneoxy)ethanol (Emulphogene

Table 4   Zinc deposition studies: Zonyl®‍FSN effect results and analysis of variance

Results   (old rating guidelines)

| Replication | Titanium Substrate | | Carbon Substrate | |
	200 mAh/cm^2	300 mAh/cm^2	200 mAh/cm^2	300 mAh/cm^2
1	8	7	5	5
2	8	8	6	6
3	9	9	8	7.5
4	8.5	7	7	6
5	9	8	8	7
6	8.5	6.5	5.5	4
7	9	8	4	4.5
8	9.5	8.5	4	3
9	8.5	7.5	6	5

BC-610) in 0.15-1.0 by weight increased cycle life was 50 to 80% at room temperature (30). The effect at 100°F was not as dramatic and the cycle life reduced at 30°F (Table 5).   In alkaline solution the main zinc species are the negatively charged zincates.

TABLE 5   Battery cycle life vs emulphogene concentration (BC 610) at several temperatures

| Concentration BC-610 % By Weight | Cycle Life | | |
	30°F 8-cell Averages	72°F 18-cell Averages	100°F 8-cell Averages
1.0	82	599 ± 35	501
0.75		509 ± 23	
0.60	82		566
0.50		613 ± 65	
0.15	166		545
Controls (2% PVA)	154	335 ± 16	468

IV. BROMINE COMPLEXATION

In the charging of a zinc/bromine battery, zinc metal and elemental bromine are produced. The bromine associates with bromide ions in solution and polybromide ions are formed. The bromine/polybromide ions must be kept separated from and stored away from the zinc electrodes to minimize self discharge.

Several chemical methods that decrease the solubility of bromine in Zn/Br_2 aqueous solutions have been reported in context with efforts to reduce the self-discharge of Zn/Br_2 cells.

The reaction between low molecular weight tetraalkylammonium halides and perchlorates with bromine-containing aqueous solutions to form sparingly soluble polyhalides has been used to lower bromine solubility in the aqueous electrolytes in Zn/Br_2 batteries (5,30).

Schemes employing polybromide ion complexation to form complexes with low aqueous solubility in the electrolyte and subsequent separation from the electrolyte, a second liquid phase, coupled with external storage of the cathode-active material, have been employed in circulating electrolyte systems. Bromine extraction that used a miscible organic cosolvent and a quaternary ammonium bromide was demonstrated by Walsh and co-workers (6). Br_2 separated from the aqueous electrolyte as a dense liquid, which comprised both the cosolvent and quaternary ammonium polybromide.

Symmetrical quaternary ammonium bromides used as single bromine-separating agents, formed solid polybromide species that were insoluble in aqueous $ZnBr_2$ solutions. These solids were electrically resistive and formed liquids at higher levels of bromination.

Unsymmetrical cyclic quaternary ammonium bromides, QBr, have relatively high aqueous solubility in concentrated (2-8M) salt solutions, are chemically stable to bromine and form liquid bromine-containing complexes at normal temperatures. The complexes

are ionic based upon specific resistance measurements.
Differential scanning calorimetry indicates no distinct liquid-
solid transitions. NMR spectrascopic and analysis indicates only
the QBr and quantitative bromine. The liquid complexes could be
viewed as bromine-fused salts.

Three unsymmetrical quaternary ammonium bromides were studied
in detail by Eustace (32). They were:

N-ethyl, N-methylmorpholinium bromide (1)

N-methoxymethyl, N-methylpiperidinium bromide (2)

N-chloromethyl, N-methylpyrrolidinium bromide (3)

They were used to prepare the electrolyte compositions given in
Table 6. The formation of the bromine-rich phase upon electrolysis
of Zn-Br$_2$ solutions containing quaternary ammonium bromides is
hypothesized to be a micellar-like separation process where pseudo-
microphases coalesce and separate from the aqueous electrolyte by
density differences. Parametric studies indicate that the critical

$$Br_2(aq) \; \ddagger \; Br_2 \text{ (fused salt)} \tag{1}$$

equilibrium is responsive to the bromide ion concentration,
temperature, and quaternary ammonium salt. This equilibrium is
influenced by the multiple ionic equilibria for Br$_2$ in the aqueous
phase of ZnBr$_2$ solutions, principally polyhalide anion formation;
for example

$$Br_2 \text{ (aq)} + Br^- \text{ (aq)} \; \ddagger \; Br_3^- \text{ (aq)}$$
$$Br_3^- \text{ (aq)} + Br_2 \text{ (aq)} \; \ddagger \; Br_5^- \text{ (aq)} \tag{2}$$
$$Br_5^- \text{ (aq)} + Br_2 \text{ (aq)} \; \ddagger \; Br_7^- \text{ (aq)}$$

and polyhalide ion pair formation, for example

$$Q^+Br^- + Br_7^- \text{ (aq)} \; \ddagger \; Q^+Br_7^- \text{ (aq)} + Br^- \text{ (aq)} \tag{3}$$

The second phase is exclusively QBr and Br$_2$ in a series of
polyhalide ion pair mixtures; for example

Table 6 Compositions and specific resistances of zinc/bromine solutions containing quaternary ammonium bromides and supporting electrolytes

Solute*	Supporting Electrolyte	(ZnBr$_2$) (M)	(Solute) (M)	(Supporting electrolyte) (M)	Density (2/cm^3)	Specific Resistance (ohm–cm)
1	—	3.0	1.0	—	1.62	22.5
1	ZnSO$_4$	3.5	1.2	—	1.67	20.9
1	KCl	3.0	1.0	0.2	1.61	21.6
1	NH$_4$Cl	3.0	1.0	3.0	1.65	8.2
1	NH$_4$Br	3.0	1.0	4.0	1.77	7.8
2	—	3.0	1.0	—	1.59	15.5
2	—	4.0	1.4	—	1.77	30.1
2	—	5.0	1.7	—	1.95	59.7
3	—	3.0	1.0	—	1.60	18.0
3	—	4.0	1.4	—	1.79	26.7
3	—	5.0	1.7	—	1.98	48.6

*The solute refers to the QBR salts in the text.

$$Q^+Br_3^- + Br_2 \; \leftrightarrows \; Q^+Br_5^-$$

$$Q^+Br_5^- + Br_2 \; \leftrightarrows \; Q^+Br_7^-$$

(4)

which is analyzed as a nonstoichiometric mix described simply as a bromine fused salt.

Increasing the concentration of bromide ion is seen to influence the position of the aqueous phase equilibria and, thus, the partitioning of bromine between the aqueous and fused salt phase

$$Q^+ Br_n(aq) \; \leftrightarrows \; Q^+Br_n^- \qquad n = 3,5,7 \qquad (5)$$

Fig. 5 shows the incremental increase of the aqueous phase Br_2 concentration found with incremental increases in the $ZnBr_2$ concentrations for N-methoxymethyl, N-methylpyrrolidinium bromide 2 solutions. Higher solubility of bromine is shown in Fig. 6 for the

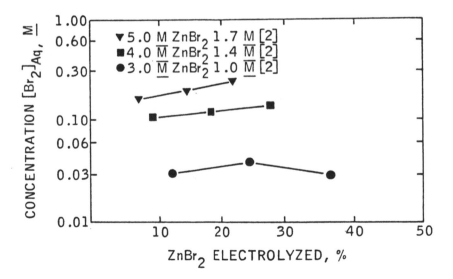

FIG. 5 *Effect of n-methyoxymethyl, methylpiperidinium bromide (2) on aqueous phase bromine concentration at 23 °C.*

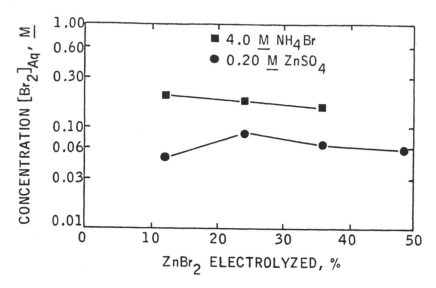

FIG. 6 *Effect of supporting electrolytes on aqueous concentration of bromine for solutions containing n-ethyl, n-methylmorpholinium bromide (1.0M) and zinc bromide (3.0M at 23°C).*

3.0M $ZnBr_2$ and 1.0M N-methyl, N-methylmorpholinium bromide 1 solution containing 4.0M NH_4Br when compared to the similar $ZnBr_2$ solution with 0.2M $ZnSO_4$. Indeed the 0.2M $ZnSO_4$ has the effect of lowering the aqueous phase bromine concentration by approximately 10% over the similar system without supporting electrolyte.

Increasing the temperature shifts the partitioning equilibria, allowing more bromine to be soluble in the aqueous phase (Fig. 7). Energies of activation for the partitioning of bromine obtained from the results of the N-ethyl, N-methylmorpholinium bromide system studies are in the range of −0.6 to 1.4 $kcal \cdot mole^{-1}$. This activation energy is lower than typical activation energies of ionic diffusion (for example, 3.8–4.6 $kcal \cdot mole^{-1}$), suggesting a solution ordering effect similar to micellar systems.

Differences between the bromine partitioning abilities of the three quaternary salts are obtained by comparing their Br_2 partition coefficients for similar composition solutions. The

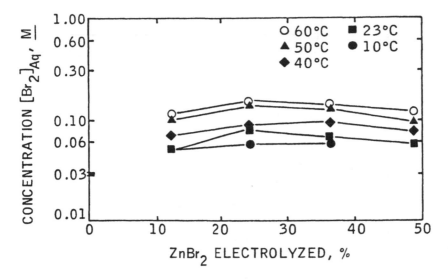

FIG. 7 *Effect of temperature on the aqueous concentration of*
bromine for solutions containing n-ethyl, n-methylmorpholinium
bromide (1). Electrolyte: 3.0M ZnBr$_2$, 1.0M (1), 0.2M ZnSO$_4$.

distribution coefficients are in the decreasing order: 3   2 > 1
(Fig. 8, 9, 10).

The circulating electrolyte zinc/bromine battery system under
development by Gould and later by Energy Research Company also uses
quaternary ammonium complexers. The chemical formulas have not
been specifically identified in reports, but patents have been
issued indicating the presence of unsymmetric alkyl quaternary
ammonium compounds. The electrolyte compositions range from 2 to 3
moles of zinc bromide and 0 to 1 moles of zinc chloride per liter
in addition 3 moles potassium chloride is present to increase
conductivity. The battery system is operated to constantly
maintain the quaternary ammonium compounds as a second phase
system. Elemental bromine is added to the system to produce a
second phase with a composition between QBr$_3$ and QBr$_4$ when the
system is at full discharge. The resulting system uses about 4 to
6 pounds of quaternary ammonium compound per kWh of stored
energy. The aqueous catholyte is admixed with the complex second

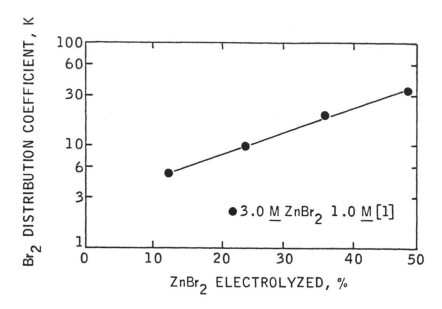

FIG. 8 *Bromine distribution coefficient vs. extent of electrolysis*
for zinc/bromide solution with n-ethyl, n-methylmorpholinium
bromide (1) at 23°C.

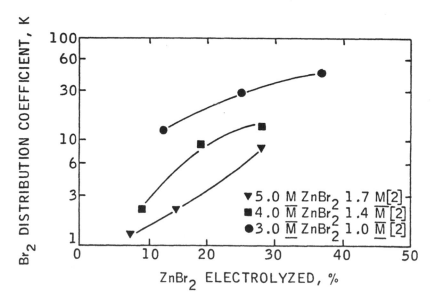

FIG. 9 *Bromine distribution coefficient vs. extent of electrolysis*
for zinc/bromide solutions containing n-methoxymethyl, n-
methylpiperidinium bromide (2) at 23°C.

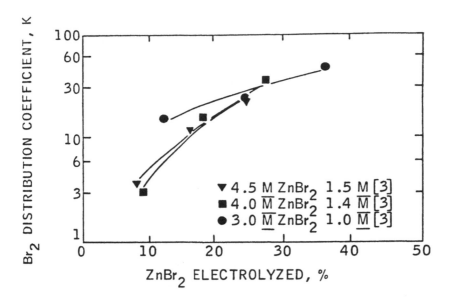

FIG. 10 *Bromine distribution coefficient vs. extent of electrolysis for zinc/bromide solutions containing n-chloromethyl, n-methylpyrrolidinium bromide (3) at 23°C.*

phase to provide an equilibrium concentration of bromine in the aqueous phase during charge and discharge. The bromine complex remains in the storage reservoir at all times and only aqueous phase bromine is circulated through the stack.

V.   CONCLUSION

Surfactants play a critical role in advanced secondary battery systems. In zinc systems, surfactants improve the plating of zinc in charge and greatly extend the lives of the batteries. The enhancement effect is evident even with small quantities. In contrast, the zinc/bromine battery systems can use several pounds of surfactant-type compounds per kWh of capacity to store bromine. The development of advanced batteries has been greatly enhanced by the use of surfactants.

REFERENCES

1. C. S. Bradley, *U.S. Patent 312,802* (1885) *409,448* (1889).

2. M. R. Bloch, *U.S. Patent 2,566,114* (1951).

3. S. Barnartt and D. A. Forejt, *J. Electrochem. Soc. 111*: 1201 (1964).

4. R. Zito, Jr., *U.S. Patent 3,382,102* (1968).

5. G. Clerici, M. DeRossi, and M. Marchetto, *Proc. of the 9th Intern. Power Sources Symp. (England)* (1974).

6. M. A. Walsh, *Proc. of the 10th Intersoc. Energy Conv. Eng. (Chicago)* (1975).

7. R. J. Bellows, *Proc. of the 12th Intern. Power Sources Symp. (England)* (1978).

8. F. G. Will, *The Zinc/Bromine Battery, Proc. of the 12th Inters. Energy Conv. Eng. Conf. (Chicago)* (1977).

9. R. A. Putt, "Assess. of Tech. and Econ. Feasibility of Zinc/Bromine Batteries for Load-Leveling", *Elec. Power Res. Inst. (Palo Alto) EM-1059* (1979).

10. A. Leo, *The 32nd Intern. Power Sources Symp. (Cherry Hill)* (1986).

11. T. Fujii, T. Fushini, T. Hashimoto, Y. Kumai, A. Hirota, H. Itoh, K. Jin-nai, Y. Kishimoto, and Kanazushi, *Proc. of the 21st Intersoc. Energy Conv. Eng. Conf. (San Diego)* (1986).

12. R. Bellows, H. Einstein, P. Grimes, E. Kantner, P. Malachesky, K. Newby, and H. Tsien, *"Dev. a Circ. Zinc/Bromine Battery, Phase I - Final Report", Contractor Report Sand 82-7022, Sandia Nat. Labs. (Albuquerque) 87185* (1983).

13. R. J. Bellows, P. Grimes, H. Einstein, E. Kantner, P. Malachesky, and K. Newby, *IEEE Trans. on Vehicular Tech., Vol. VT-32, No. 1 (1983).*

14. *R. J. Bellows, C. Elspass, H. Einstein, P. Grimes, E. Kantner, P. Malachesky, and K. Newby, Proc. of the 18th Inters. Energy Conv. Eng. Conf. (Orlando)* (1983).

15. R. Bellows, H. Einstein, P. Grimes, E. Kantner, P. Malachesky, K. Newby, and H. Tsien, *"Dev. of a Circ. Zinc/Bromine Battery; Phase II - Final Report", Contractor Report Sand 83-7108; Sandia Nat. Labs., (Albuquerque) 87185* (1983).

16. R. Bellows, P. Grimes, and P. Malachesky, *"Zinc/Bromine Battery System Tech.", Proc. Elec. and Hybrid Vehicle Assess. Seminar (Gainesville)* (1984).

17. P. Grimes and R. Bellows, *Proc. of the 19th Inters. Energy Conv. Eng. Conf. (San Francisco)* (1984).

18. R. Bellows, H. Einstein, E. Kantner, P. Grimes, and P. Malachesky, *Proc. of the 20th Inters. Energy Conv. Eng. Conf. (Miami Beach)* (1985).

19. P. Grimes, R. Bellows, and M. Zahn, *"Shunt Current Control in Electrochem. Systems - Theoretical Analysis"*, *Electrochem. Cell Design*, R. White, Editor, (New York) (1984).

20. P. Grimes and R. Bellows, *"Shunt Current Control in Electrochem. Systems - Appl.*, *Electrochem. Cell Design*, R. White, Editor, (New York) (1984).

21. R. D. Naybour, *J. Electrochem. Soc. 116:* 520 (1969).

22. D. J. MacKinnon and J. M. Brannen, *J. Appl. Electrochem., 12:* 21 (1982).

23. J. W. Diggle and A. Damjanovic, *J. Electrochem. Soc., 117:* 65 (1970).

24. D. C. Symons, *Soc. of Automotive Engineers Trans., No. 730253* (1973).

25. D. C. Symons and H. J. Hammond, *Elec. Power Res. Inst. No. EM-249, Energy Dev. Assoc. (Madison Heights)* (1976).

26. R. A. Putt and A. Attia, *"Dev. of Zinc/Bromide Batteries for Stationary Energy Storage"*, *Elec. Power Res. Inst. (Palo Alto)*

27. R. A. Putt, A. J., Attia, L. Po-Yen, and J. H. Yeland, *"Dev. of Zinc/Bromine Batteries for Utility Energy Storage"*, *Elec. Power Res. Inst. (Palo Alto) EM1717* (1981).

28. F. G. Will and F. F. Holub, *U.S. Patent 4,040,916* (1977).

29. E. C. Meidensha, *U.S. Patent 4,510,218* (1985).

30. F. Rallo and P. Silvestroni, *J. Electrochem. Soc., 119:* 1471 (1972).

31. D. Eustace, *J. Electrochem. Soc., 127:* 528 (1980).

Non-Conventional Energy Production
Group Discussion I

Darsh T. Wasan, Illinois Institute of Technology, rapporteur

The discussion was concerned largely with the role of surfactants in coal/water slurries as direct fuel substitutes and
multiphase interfacial phenomena in fuel cells and batteries. The
highlights of this discussion are as follows:

I. Coal/Water Mixtures as Direct Fuel Substitutes

The preparation of such a suitable mixture requires dispersants or surfactants that should result in high solid loading,
low viscosity and high stability of the dispersions. It was
noted that commonly used dispersants such as naphthalene sulfonates are expensive and the lignosulfonates which are less
expensive generally age poorly, therefore there is a need for
synthesizing surfactants to optimize the viscosity and stability
of the desired coal/water slurry systems. Also, a need exists
for understanding the basic mechanisms of the adsorption of surfactants on coal particles, since these mechanisms are not well
understood. Furthermore, the relationship between the slurria-
bility of coal particles and the rank of coal needs to be better
understood. For example, it is easier to slurry high ranking coal
such as anthracite, which contains a lesser amount of oxygenated

compounds than low ranking coal such as lignite, which has more
oxygenated compounds.

II.   Energy Conversion Devices

The use of surfactants as additives in metal plating was
recognized and the research needs for understanding the mechanisms
of interactions between surfactants and electrode surfaces were
pointed out. Absence of contaminants that might act as poisons
for the reactions involved and stability to oxidation and reduction
at the anode and cathode, respectively, are requirements for the
surfactants used. Mechanisms need to be understood in order to
achieve high current density without shorting of the electrode in
battery systems. An urgent need was also felt for understanding
the multiphase interfacial phenomena involved in the use of surfactants to form aqueous teflon emulsions for porous gas-phase
electrodes.

Non-Conventional Energy Production
Group Discussion II

Graham Barker, Witco Chemical Corp., rapporteur

Q: If proven commercially feasible, how far are you from commercial production of the Zn-Br$_2$ battery?

A: Two and a half to three years. The use of electric vehicles
may return because of high pollution caused by current gas and
oil vehicles. Technical feasibility is here for Zn-AgO batteries
but they are not economical and they are not commercially attractive. Lead-acid batteries are quite heavy and do not have the
range but the Zn-Br$_2$ batteries are all plastic, realatively cheap,
and have a greater range.

Q: If accidents occur, what are the safety considerations of the
Zn-Br$_2$ battery?

A: In case of accidents, tests have shown that the Zn-Br$_2$ batteries are safer than the chlorine-based batteries which are in turn
safer than lead-acid batteries. This is because of the contain-
ment of the bromine.

Q: How far can a battery be discharged before it starts to lose
efficiency?

A: About half way before loss of efficiency occurs. With non-
uniformity of plating and after 50-100 cycles, we have to stop
and start over.

Q: Have you conducted a systematic study of surfactants for the
Zn-Br$_2$ battery?

A: We have looked at many families of compounds but there was no
intensive systematic study of surfactants. The tetrabutyl ammonium quaternary appeared best for plating efficiency, but it got
chewed up by the bromine.

Q: How is the battery activated, for instance, in the case of
starting an automobile?

A: If the battery is at potential, then one can use the main
battery, otherwise we would have to use a supplemental battery as

Q: I've used Zn electrodes in the past and I wonder how
dendritic growth on the electrode and overall roughness affects
performance?

A: Surface roughness greater than 3-4 times the geometric surface is not practical.

Q: Have you considered electroless plating?

A: I do not believe it would be practical nor applicable here.

Q: Can you comment on other non-conventional energy sources
currently being researched?

A: In the battery area, the sodium-sulfur battery operating at
300 degrees C. is still being investigated even though both Ford
and GE have looked into that technology and stopped their
research efforts. This has been going on for at least 20 years.
There are current research efforts on the Zn-Cl$_2$ battery at ambi-
ent temperatures and the Ni-Zn battery. Each has its own advantages and disadvantages. Work is still continuing on these
systems.

Q: Does the price of oil have any impact on research on non-conventional energy sources:

A: No. The battery is used as a storage device only and is completely independent of oil prices.

Q: What is the conventional range of storage batteries in
electric vehicles?

A: Depending on speed, the lead-acid will go 35-40 miles. Other more advanced batteries than we have discussed will go about 100-125 miles. The aluminum-air battery would have a range of several hundred miles. One can equip a vehicle to go further, but more batteries would be required. Between the weight and expense you reach a point of diminishing returns.

Q: Has any systematic study been undertaken on the effect of surfactants, such as structure/function relationships?

A: Although the proper surfactant is critical in the plating operation, we are not sure how they work and not thoroughly familiar with the mechanism. At present there is some science in the selection of a surfactant for a particular system and there are optimum surfactants for each battery sytem. Each electrolyte has its own characteristics and the electrode will also dictate the selection of the appropriate surfactant. There is no specific class that works best in all cases.

Q: Are commercial plating additives of any practical use in your application?

A: No. Most formulations are proprietary and little is divulged about the composition. In addition, these additives are used in plating applications for decorative purposes. In our case, we are concerned with the stability of the system in the oxidation and reduction cycling occuring in the battery.

6
Surfactants and Biotechnology

Saul L. Neidleman

Cetus Corporation
Emeryville, California

I. INTRODUCTION

Biological systems learned that molecules containing a hydrophilic and a hydrophobic domain provide useful properties long before the commercial utility of their chemically synthesized cousins was uncovered by industry. Such compounds are surface active agents affecting the physical characteristics of interfaces between immiscible phases. For example, these chemicals can reduce the surface tension at air/water interfaces or the interfacial tension at oil/water interfaces. There are other boundaries at which surface active agents interact including the liquid/solid interface, where the solid may be a microorganism and surface wetting may be facilitated.

The surface active chemicals may be divided into two major groups: biosurfactants and bioemulsifiers. Simplistically, biosurfactants favor emulsification of oil in water, while bioemulsifiers stabilize this emulsion. In contrast to biosurfactants, bioemulsifiers generally do not reduce

interfacial tension but prevent oil droplets from coalescing.
There is often confusion in the literature over this distinction
and the term biosurfactant is used to cover all situations, so
that the utility of biosurfactants has been stated to extend
from emulsification and demulsification through wetting,
detergency, and foaming (1).

The broad range of biological systems from bacteria to
humans, associated with their variety of environmental and
physiological demands, has resulted in many biosurfactant
requirements and a great diversity of biochemical structures to
satisfy these needs (1-7). What are some of the natural roles
that have been suggested for these surface active chemicals?

In some cases, the functions of biosurfactants resemble
those of commercial surfactants, favoring adherence to food
particles or growth surfaces and emulsification of water-
insoluble food substances. The bacterium *Acinetobacter
calcoaceticus* RAG-1, adheres to droplets of n-hexadecane and
also produces the emulsifier, emulsan. A spontaneous mutant MR-
481 retains the ability to synthesize normal amounts of emulsan
but has lost the capacity to adhere to n-hexadecane. While RAG-
1 can grow on n-hexadecane, MR-481 can not, except in the
presence of additional emulsan. The added emulsan facilitates
increased emulsification of n-hexadecane necessary for MR-481
growth in the absence of the surface active adherence factor,
whereas with RAG-1 that has the adherence factor, additional
emulsan was not needed. Growth of this bacterium on n-
hexadecane is, then, a balance of adherence and level of
emulsification mediated through surface active agents (7-11).
Strains of *Pseudomonas aeruginosa* have shown a requirement for
an extracellular rhamnolipid to facilitate hydrocarbon
utilization. Rhamnolipid negative mutants grew poorly unless
supplemented with rhamnolipid from the parent strain (12).

In other cases, the interfacial effects of the
biosurfactants are accompanied by additional biological
properties such as antibiotic activity against microorganisms
and tumor cells (13). It is not clear what the relationship of

such growth inhibiting activity of surface active agents has to
the natural function of such compounds. Among the most active
of these compounds is 6,6'-dimycoloyl-α-trehalose, produced by
Mycobacterium tuberculosis.

The purposes of this paper are to briefly indicate the
wide chemical spectrum of biosurfactants, the new possibilities
for enzymatic synthesis of surfactants, and to consider other
biotechnological approaches for producing novel biosurfactants
not presently available in nature or industry.

II.   CHEMICAL VARIATION IN BIOSURFACTANTS

The structures of biosurfactants produced by microorganisms give
some concept of the chemical variability seen in nature. Table
1 lists a few characteristic compounds. The substances can be
divided into five major groups: glycolipids, neutral lipids and
fatty acids, lipopeptides and lipoamino acids, phospholipids,
and polymers (1-7). The two lipopeptides, surfactin and
polymixin, are examples of surfactant antibiotics, illustrating
that surfactants may have "secondary" properties. The examples
shown clearly demonstrate the extraordinary diversity of
biosurfactant chemistry. It is fascinating to see the broad
range of biochemicals that can offer the functional attributes
needed to satisfy the requirements of specific systems within
living cells.

III. ENZYMATIC SYNTHESIS OF CARBOHYDRATE ESTERS

One group of commercially available surfactants has a close
structural similarity to certain biosurfactants. These are the
carbohydrate esters, such as sucrose esters, which have a wide
range of applications in the food, cosmetic, pharmaceutical, and
detergent industries (14). These include uses as low calorie
fat substitutes, bulking agents, emulsifiers, and surfactants.

The production of carbohydrate esters has traditionally been through chemical synthesis. The methods have problems including color development and lack of specificity. There are eight esterifiable hydroxyl groups on sucrose, and the products obtained are generally an ester mix. The structure of a sucrose monoester is shown in Figure 1. It can be seen that seven additional - OH groups are available for esterification.

The numbers of ester linkages determines, in large measure, the properties of the product. Derivatives with three or less fatty acids are surfactants with interest as food additives because of their emulsifying, stabilizing, and conditioning properties. Derivatives esterified with six to eight fatty acids are fatlike and nonabsorbable with potential as fat substitutes.

Until very recently, the enzymatic synthesis of carbohydrate esters with purified enzymes was unreported. Japanese workers have now taken the first important step: they have enzymatically prepared esters of sucrose, glucose, fructose, and sorbitol with oleic, linoleic and, poorly, with stearic acid (15). The enzyme yielding the best results was the lipase of *Candida cylindracea*. Reactions were run under aqueous

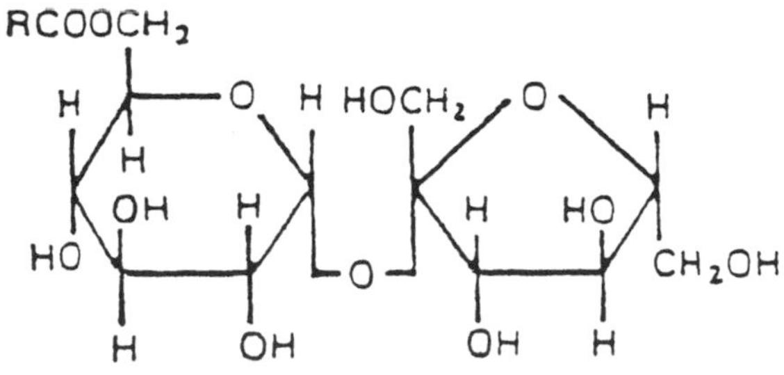

$RCOOCH_2$ = ester linkage with fatty acid

FIG. 1 *Sucrose monoester*

conditions at 40°C, pH 5.4 for 72 hours. Conversions up to 68% were obtained.

This important discovery opens a new field of research for the synthesis of novel surfactants under mild reaction conditions and with the advantage of specificity offered by many enzymes. There are many different lipases with varying requirements for substrate structure and pH and temperature optima.

The similarity in structure between these synthetic carbohydrate esters and the glycolipids illustrated in Table 1 strongly indicates that nature is replete with enzymes for the production of novel surfactants. These natural glycolipids appear to function as highly specific detergents since their ability to exhibit growth inhibiting activity is very structure dependent and has no simple explanation (13). Diesters of trehalose, for example, have higher activity than monoesters of methyl glucoside and the occurrence of an axial hydroxyl group in mannoside, alloside, and galactoside derivatives also reduced the toxic properties of these acyl-sugars. All that can be said is that the balance between the hydrophobic and the hydrophilic parts of the molecule determines the physical and biological properties of these molecules. The unusual structures of these and other biosurfactants and their range of physical and biological properties should aid surfactant chemists in conceiving and synthesizing new synthetic surfactants for commercial applications.

IV. LIPID MODIFICATION IN BIOSURFACTANTS

In the case of the carbohydrate esters, the lipid moiety can be varied as part of the procedures of chemical synthesis. The situation in biosynthesis is similar. Lipids are a frequent constituent of biosurfactants and their structure may be

TABLE 1 *Biosurfactants from microorganisms*

Class	Typical Structures	Producer
1. Glycolipids:		
Trehalose Lipids	Trehalose esterified with 2 α-branched-β-hydroxy carboxylic acids such as 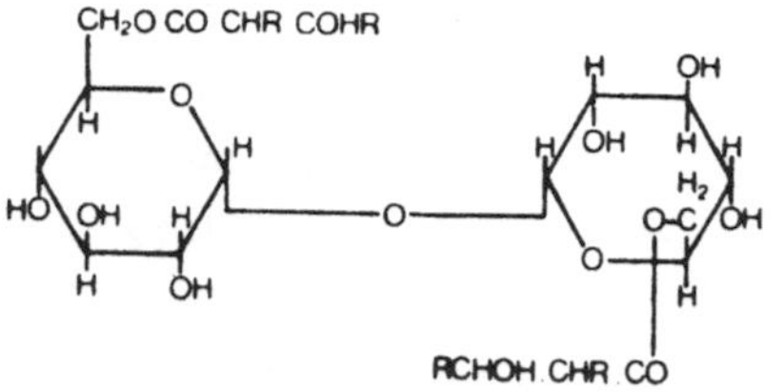	Arthrobacter, Mycobacteria, Nocardia, Corynebacteria, Rhodococcus

CH₂O CO CHR COHR

RCHOH CHR CO

| Rhamnolipids | 2 β-hydroxycarboxylic acids glycosidically linked to rhamnose such as 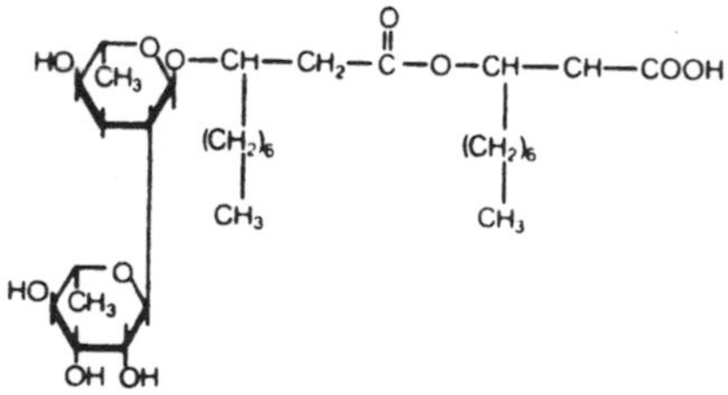 | Pseudomonas |

| Sophorolipids | (ω-1)-hydroxycarboxylic acid glycosidically linked to sophorose (may be lactonized) such as 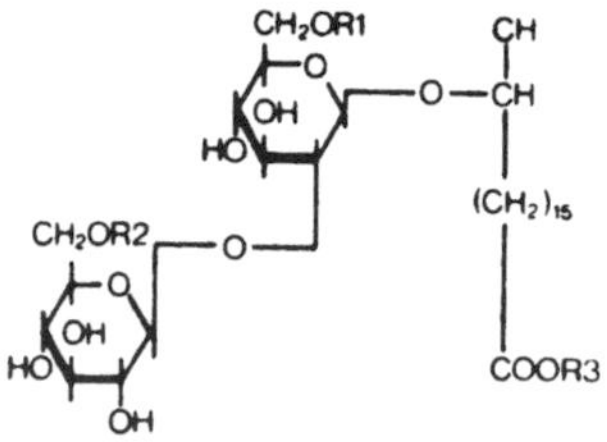 | Torulopsis |

| 2. Neutral Lipids/ fatty acids | Glycerides, fatty acids, fatty alcohols, wax esters | Acinetobacter, Clostridia |

TABLE 1 (Cont'd) *Biosurfactants from microorganisms*

Class	Typical Structures	Producer
3. Lipopeptides Lipoaminoacids	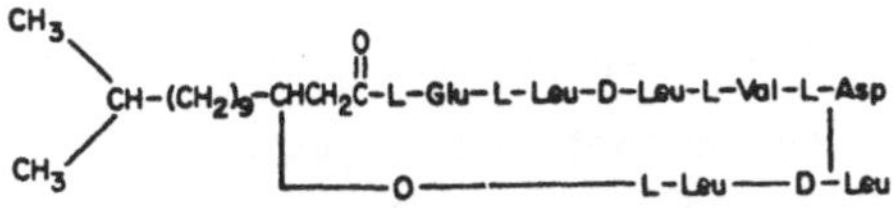 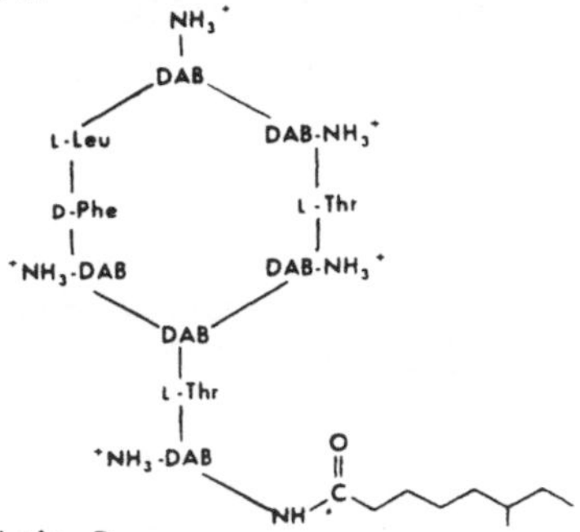 fatty acid acylated amino acids or cyic peptides such as	Bacilli, Nocardia, Corynebacteria, Streptomyces, Mycobacteria, Pseudomonas, Agrobacteria, Gluconobacter

CH₃
CH-(CH₂)₉-CHCH₂C-L-Glu-L-Leu-D-Leu-L-Val-L-Asp
CH₃
└──────O────────L-Leu──D-Leu

Surfactin

NH₃⁺
|
DAB
|
L-Leu   DAB-NH₃⁺
| |
D-Phe   L-Thr
| |
⁺NH₃-DAB   DAB-NH₃⁺
| |
DAB
|
L-Thr
|
⁺NH₃-DAB ── NH-C-...

Polymixin B
(DAB = α, γ-diaminobutyrate
[NH₂CH₂CH₂CH(NH₂)COOH]. Aliphatic
residue is 6-methyloctanoic acid.)

| 4. Phospholipids | | Rhodococcus Thiobacillus |

O
||
R—C—O—CH₂
R'—C—O—CH   O
O   H₂C—O—P—O—CH₂—CH₂—NH₃⁺
|
O⁻

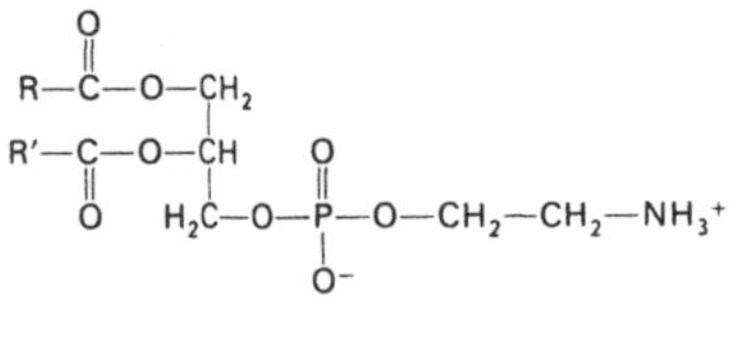

phosphatidyl ethanolamine

5. polymers	lipoheteropolysaccharide such as emulsan	Acinetobacter
	polysaccharide-lipid complex	Candida
	polysaccharide-protein complex	Acinetobacter

modified by growing the producing microorganism under varying
conditions.

   Bacterial isolate H-13A grown on C-13 to C-16 alkanes
produced an extracellular glycolipid surfactant with altered
fatty acid composition depending upon the carbon number of the
alkane substrate (16). The qualitative and quantitative
variations in fatty acids did not appear to directly relate to
the carbon number of the substrate. This indicated that the
fatty acids of the surfactant were synthesized *de novo* in the
fatty acid pathway in a physiological response to the surfactant
properties required for efficient metabolism of each, individual
substrate. This is an impressive example of the exquisite

TABLE 2 *Effect of alkane carbon number on wax ester formation
by Acinetobacter sp HO1-N*

Substrate	→	Fatty Acids	+	Fatty Alcohols	→	Wax Esters		
n-Hexadecane		16:0, 16:1[a]		16:0, 16:1		32:0		
						32:1		
						32:2		
n-Eicosane		16:0, 16:1		20:0, 20:1		36:0	38:0	40:0
		18:0, 18:1				36:1	38:1	40:1
		20:0, 20:1				36:2	38:2	40:2

[a]16, 18, 20, 32, 34, 36, 38, 40 = number of carbon atoms.
 0, 1, 2, = number of double bonds.

flexibility and sensitivity of biological systems to the range
of properties needed in surface active agents involved in making
foodstuffs available to living systems.

Further evidence for the responsiveness of biological
systems to available nutrients and the need for surface active
agent synthesis can be cited. Maximum yields of glycolipid
surfactants were produced by the yeast *Torulopsis bombicola* in
the presence of a mixture of glucose and vegetable oil (17).
Whereas most biosurfactants have been produced in the presence
of hydrocarbons, the substance surfactin was produced by
Bacillus subtilis with glucose as the substrate and hydrocarbons
surpressed production of the substance (18).

In contrast to the lack of relation between the carbon number of
the substrate and the fatty acid carbon number in the
biosurfactants produced by H-13A, the neutral lipid, wax ester
bioemulsifiers of *Acinetobacter* sp. HO1-N were directly
responsive to the carbon number of the alkane substrate as shown
in Table 2. Further, the level of unsaturation in the wax
esters was clearly affected by the growth temperature. Table 3
shows that unsaturation increased with decreasing growth
temperature (19) with either N-eicosane or ethanol as
substrate. In other words, the wax esters became more
hydrophilic with decreasing temperature.

A similar effect of temperature was reported in
rhamnolipid surfactant production by *Pseudomonas* sp. DSM2874
(20). The results are shown in Table 4. Lower growth
temperature resulted in the biosynthesis of more hydrophilic
surfactants in addition to the more hydrophobic surfactants seen
as higher temperature. These few examples reflect the
flexibility of biological systems in producing surfactants
specific to a particular environmental situation.

TABLE 3 *Effect of temperature on unsaturation of wax esters*
produced from ethanol and N-eicosane by Acinetobacter
sp. HO1-N

A. Ethanol

Temperature	Di-Ene fractions (32:2, 34:2, 36:2) Total	Mono-Ene fractions (32:1, 34:1, 36:1) Total	Saturated fractions (32:0, 34:0, 36:0) Total
17 C	72%	18%	10%
24 C	29%	40%	31%
30 C	9%	25%	66%

B. N-Eicosane

Temperature	Di-Ene fractions (36:2, 38:2, 40:2) Total	Mono-Ene fractions (36:1, 38:1, 40:1) Total	Saturated fractions (36:0, 38:0, 40:0) Total
17 C	71%	24%	5%
24 C	41%	37%	22%
30 C	35%	40%	25%

V.   GENETIC ENGINEERING AND BIOSURFACTANTS

Genetic engineering will also have its impact on
biosurfactants.   Pulmonary surfactant is a phospholipid -
protein complex that lowers surface tension at the air-liquid
interface in the alveoli of the mammalian lung. It is necessary
for normal respiration. Frequently, premature infants have a

TABLE 4   *Rhamnolipids produced by resting cells of* <u>*Pseudomonas*</u>
          sp. DSM 2874

Cultivation Conditions	% Total Rhamnolipid			
	R1	R2	R3	R4
Resting cells	42	15	41	2
C-source: n-alkanes				
Temperature: 30°C				
(168 h)				
Resting cells	57	--	43	--
C-source: n-alkanes				
Temperature: 37°C				
(168 h)				

R1: R = $-CH-CH_2-COOH$
          $(CH_2)_6$
          CH_3

R2: R = H

R3: R = $-CH-CH_2-COOH$
          $(CH_2)_6$
          CH_3

R4: R = H

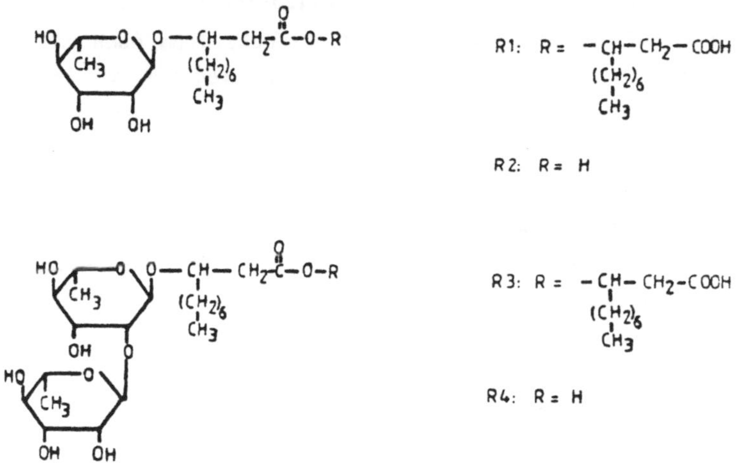

deficiency of pulmonary surfactant and suffer respiratory failure. In all species, the surfactant is mainly composed of two major protein molecules of relative mass 10K and 32K and dipalmitoylphosphatidylcholine. The human gene for the 32K protein has been cloned and studies on its expression are proceeding (21). One goal for this research could be the preparation of large quantities of the 32K protein to be used in the treatment of insufficiency of pulmonary respiration.

VI. CONCLUSION

It is clear that, in addition to their function in areas of commercial interest, surfactants play a major role in the effective accomplishment of life processes, such as adherence and food emulsification in microorganisms and surface tension in the mammalian lung. The two areas of application draw upon the surface activity of these compounds, and so data from each may make a contribution towards understanding or progress in the other. The chemical and physical requirements for commercially useful surfactants need to be defined in terms translatable into biochemical structures that may differ markedly from those of traditional chemical surfactants. On the other hand, known biochemical surfactants may offer some novel ideas to the surfactant chemist for new and useful compounds and, in many instances, because of the pressures of product cost, this may be the biosurfactant's only contribution, particularly in commodity uses. For specialty uses, such as medical applications, biosurfactants may have a cost advantage over chemical surfactants. This could result because of highly specific and difficult synthetic steps necessary to prepare the chemical surfactants.

The antibiotic field offers a valuable model for the surfactant field. The β-lactam antibiotics, including the penicillins, cephalosporins, and monobactams, are dominant compounds in present use. They are all based on β-lactam-containing ring systems derived from natural products. Chemical synthesis can not compete. Nature offered the prototype molecules and chemists then exploited these to yield the compounds of commerce. It is reasonable to state that the chemists would not have, even to this day, randomly or semi-rationally, synthesized β-lactam antibiotics by chemical means. Perhaps, by analogy, super-surfactant prototypes exist in nature, awaiting discovery and exploitation by chemists.

REFERENCES

1. Gerson, D. F. and Zajic, J. E. Microbial Biosurfactants. *Process Biochem. 14:* 20-22, 29 (1979).

2. Zajic, J. E. and Panchal, C. J. Bioemulsifiers. *CRC Crit. Rev. Microbiol. 5:* 39-107 (1976).

3. Zajic, J. E. and Seffens, W. Biosurfactants. *CRC Crit. Rev. Biotechnol. 1:* 87-105 (1984).

4. Parkinson, M. Biosurfactants. *Biotech. Adv. 3:* 65-83 (1985).

5. Cooper, D. G. Biosurfactants. *Microbiol. Sci. 3:* 145-149 (1986).

6. Hayes, M. E., Nestaas, E., and Hrebenar, K. R. Microbial Surfactants. Chemtech, pp. 239-243 (April 1986).

7. Rosenberg, E. Microbial Surfactants. *CRC Crit. Rev. Biotechnol. 3:* 109-132 (1986).

8. Rosenberg, E., Zuckerberg, A., Rubinovits, C., and Gutnick, D. L. Emulsifier of *Arthrobacter* RAG-1: Isolation and Emulsifying Properties. *Appl. Environ. Microbiol. 37:* 402-408 (1979).

9. Rosenberg, E., Perry, A., Gibson, D. T., and Gutnick, D. L. Emulsifier of *Arthrobacter* RAG-1: Specificity of Hydrocarbon Substrate. *Appl. Environ. Microbiol. 37:* 409-413 (1979).

10. Rosenberg, M. and Rosenberg, E. Role of Adherence in Growth of *Acinetobacter calcoaceticus* RAG-1 on Hexadecane. *J. Bact. 148:* 51-57 (1981).

11. Neufeld, R. J. and Zajic, J. E. The Surface Activity of
    Acinetobacter calcoaceticus sp. 2CA2. *Biotech. Bioeng. 26:*
    1108-1113 (1984).

12. Itoh, S. and Suzuki, T. Effect of Rhamnolipids on Growth
    of *Pseudomonas aeruginosa.* Mutant Deficient in n-Paraffin-
    Utilizing Ability. *Agric. Biol. Chem. 36:* 2233-2235
    (1972).

13. Asselineau, C. and Asselineau, J. Trehalose-Containing
    Glycolipids. *Prog. Chem. Fats Other Lipids 16:* 59-99
    (1978).

14. Walker, C. E. Food Applications of Sucrose Esters. *Cereal
    Foods World 29:* 286-289 (1984).

15. Seino, H., Uchibori, T., Nishitani, T., and Inamasu, S.
    Enzymatic Synthesis of Carbohydrate Esters of Fatty Acids
    (1) Esterification of Sucrose, Glucose, Fructose, and
    Sorbitol. *J. Am. Oil Chem. Soc. 61:* 1761-1765 (1984).

16. Finnerty, W. R. and Singer, M. E. A Microbial Surfactant -
    Physiology, Biochemistry, and Applications. *Dev. Ind.
    Micro. 25:* 31-40 (1984).

17. Cooper, D. G. and Paddock, D. A. Production of a
    Biosurfactant from *Torulopsis bombicola.* *Appl. Exp.
    Microbiol. 47:* 173-176 (1984).

18. Cooper, D. G. Unusual Aspects of Biosurfactant Production
    in Biotechnology for the Oil and Fat Industry (C. Ratledge,
    P. Dawson, and J. Rattray, eds.), Chapter 24, pp. 281-287
    (1984).

19. Neidleman, S. L. and Geigert, J. Biotechnology and
    Oleochemicals: Changing Patterns. *J. Am. Oil Chem. Soc.
    61:* 290-297 (1984).

20. Syldatk, C., Lang, S., Matulovic, U., and Wagner, F.
    Production of Four Interfactial Active Rhamnolipids from n-
    Alkanes or Glycerol by Resting Cells of *Pseudomonas species*
    DSM2874. *Z. Naturforsch. C. 40:* 61-67 (1985).

21. White, R. T., Damm, D., Miller, J., Spratt, K., Schilling,
    J., Hawgood, S., Benson, B., and Cordell, B. Isolation and
    Characterization of the Human Pulmonary Surfactant
    Apoprotein Gene. *Nature 317:* 361-363 (1985).

Biotechnology
Group Discussion I

Robert B. Login, GAF Corp., rapporteur

This discussion was concerned with the potentialities of biotechnology for the synthesis of organic molecules, including biosurfactants. The microbial synthesis of enzymes, such as lipases and proteases, for use in detergents was also discussed, including the achievement of bleach stability through the removal of methionine. In contrast those attainable by separative procedures from natural products, which may vary with the nature of the raw material, biosurfactants produced by biotechnological processes can be made with fully reproducible properties.

Purified enzymes isolated from naturally occurring organisms can be used as is or may be biologically or chemically modified according to requirements. Their solubility characteristics can be changed by the incorporation of hydrophobic (lypophilic) moieties. This allows their use in non aqueous solvents and this often permits their use at much higher temperatures since, in the presence of only small amounts of water, thermal stability is increased. Hydrolytic enzymes, in the presence of small amounts of water, may also catalyze condensation reactions.

Of immediate importance is the enzymatic syntheses of sugar esters. This class of surfactants is receiving renewed interest in

cosmetics because of its mildness to skin and eyes. At present,
these esters are manufactured by conventional chemistry; however
enzymatic production under mild and simple conditions could offer
new classes of economically-attractive compounds.

There are a number of positive aspects to the utilization of
enzymes for surfactant synthesis or manufacture: 1) relatively
pure or easily purified products can be prepared,2) syntheses can
be achieved in aqueous or nonaqueous medium, at high or low temperature, depending upon the modification of the enzyme (e.g., through
derivatization with lipophilic groups), 3) enzymes can be used to
catalyze a reaction in both directions, e.e., esterification or
hydrolysis, depending upon the amount of water present, 4) reaction
can occur on specific sites without the need for protecting groups,
and 5) enzymatically-produced biosurfactants can suggest new types
of surfactants, not previously conceived of, for synthesis by
conventional methods.

Biotechnology
Group Discussion II

Graham Barker, Witco Chemical Corp., rapporteur

Q: Is your company trying to market or license some of the new technologies developed utilising lipases and other enzymes?

A: At the beginning my company started looking for joint ventures. The first problem we looked into was the reason for the apparent increase in viscosity that occurred with certain hydrocarbon oils and why there was an apparent increase in the length of the hydrocarbon chain lengths. We obtained some crude oil and developed biowax. (The very viscous dark material usually found on beaches when there is an oil spill) The interesting aspect was that there was a possibility that two C_{20} hydrocarbons could be stitched together to yield a C_{40} hydrocarbon. However, it appeared that the bacteria involved use up the shorter chains first. After the shorter chains are used up, the solubility characteristics of the oil changes because the higher alkane chain lengths precipitate out and you get these balls comprised of the higher chain alkanes. Thus there was no stitching, just a change in the crude oil composition. Still investigating the stitching reaction, we took a C_{16} and wanted to see if a C_{32} could be synthesised. Analysis of the final composition using GC, indicated that no C_{32} alkane was formed. Further analysis showed that a C_{32} ester wax was formed.

Initial results indicated that the C_{32} esters were saturated. At
this time, we were asked if we could make jojoba oil, an unsaturated long chain ester. We could separate wax esters based on the
amount of unsaturation. A method was developed and in fact a
jojoba oil could be made and also cheaper than could be made via
conventional means. Jojoba is a mixture of three wax esters and
in engine tests it appeared better than conventional lube oils.
We also looked into the preparation of sperm oil.

If a crude oil were taken as the starting point, and it con-
tained C_{14}, etc., and C_{10} was added so a mix could be obtained, wax
esters of C_{10} with all the other alkanes were prepared. At lower
temperatures one can make the unsaturated esters. We spent about
5 years on the use of biotechnology for tertiary oil recovery. We
were requested to develop at that time a method to convert oil to
methane. This was nothing special since the literature was full of
references on this technology. However, if one could use the
ability of a microorganism to move across gradients toward a
specific concentration, many attractive opportunities became
apparent since the bacteria would be attracted towards a specific
oil. This could be demonstrated with an agar plate. Nothing of
commercial importance developed out of this technology.

Q: Was any Company interested in exploring the synthetic Jojoba
oil route?

A: Amoco showed no further interest in this particular technology
at that time. Battelle Research Institute conducted a market
analysis that did not encourage further expoloitation. I believe
that the general area of making esters of glycerine and other
alcohols or polyols using this technology will explode into a
real commercial contender for the preparation of new and novel
esters. For instance, it has been demonstrated that,despite
popular belief, enzymes can work very well in solvent systems.
Lipases and esterases that normally hydrolyze fats and other esters,
can be used to synthesize esters in the presence of as little as
0.1% water. This would be ideal for the preparation of certain
difficult-to-synthesize fragrance oils. There are specific

lipases that could direct the fatty group to a certain hydroxy group of a polyol such as glycerine. Then again, there are non-specific esterases or lipases that could be utilised where non-specificity can be tolerated. The technology is now poised to develop new types of surfactants such as sugar esters and esters of other carbohydrates that have not been exploited.

Q: Can systems be developed to convert saturated to unsaturated esters?

A: To date we haven't seen data to substantiate this and no organisms to my mind are that specific. Glycerides fall into this category and the degree of unsaturation depends on the temperature.

Q: Have you worked with certain carboxylic acids to prepare specific esters?

A: We have never tried to optimize any systems. The initial work with the wax esters used about 0.1% of the microorganism. Some yields are quite low (even lower than 1%), since the organism uses the ester that is formed. However, there are certain mutants that can be used to increase yields of the desired product. These mutants cannot use the higher esters, such as those derived from hexadecane, and yields as high as 30% can be obtained. It may be possible to obtain efficiencies close to theoretical.

Q: Are the new experimental Proctor and Gamble polyesters that are not metabolized derived via biotechnology?

A: I do not believe it is an enzymatic system.

Q: Have there been any commercial successes of bioderived products?

A: One product that comes to mind are the Petroferm products. I have heard that Emulsan, one of their bioderived surfactants, can stabilize a 70% solids coal slurry. The main drawback may be economic. I have not heard of other commercial succeses but I believe the technology is all in place to develop some remarkable new systems for new unique surfactants and esters. An attempt is being made to develop bleach-stable enzymes that can be used in detergent systems. If the methionine is removed from the enzyme and replaced with alanine, for instance, a bleach-stable enzyme will be formed.

Q: Has anyone else looked into the jojoba oil project?

A: Since Amoco did not have any further interest in the project, there was no further commercial interest. If one is interested in pursuing this, there are certain personnel that can be contacted for further details. Although the price may be high, the wax esters can be modified to give a different wax ester mix that would be a jojoba oil analogue with special properties. Using ethanol and acetic acid, one could get all the even carbon number was esters but with propionic acid one would get a whole range of esters, such as C_{31}, C_{32}, C_{33}, etc.

Q: How long a chain can one build using this technology?

A: We cannot answer that question since there is no present specific need for this. Some bacteria can make C_{60} fatty acids but no specific projects to address that problem have been undertaken.

Q: Do you think that some of this technology can impact on some of the commodity-type of surfactants?

A: In general, biologically-derived products are more expensive than chemically-derived products and it is doubtful if there would ever by any significant impact on commercial surfactants. With certain lipases, it is possible that with 80-90% yields one can get unique surfactants with a good possibility of commercial success.

7
Surfactants in Novel Separation Techniques

Donald B. Wetlaufer

Chemistry Department
University of Delaware
Newark, Delaware

The purpose of this paper is to outline some relatively recent
developments in the use of aqueous solutions of surfactants for
chemical separations. Most of this work has been done on the
analytical scale, but in some cases preparative scale separations
appear to be possible.

We will here discuss micellar-enhanced ultrafiltration,
micellar chromatography, the sodium dodecyl sulfate gel
electrophoresis of proteins, and submicellar surfactant-mediated
protein chromatography. The analytical use of detergent micelles
to enhance sensitivity in luminescence detection (1), to improve
yields in derivatization, and in electroanalytical chemistry (2)
can only be mentioned in passing. Similarly, the foam
fractionation of proteins (3), which depends on the intrinsic
surface activity of these large molecules, will not be treated.

Conventional ultrafiltration does not remove low molecular
weight organic molecules from water. The addition of surfactant
at concentrations substantially above the critical micelle
concentration (CMC) results in the transfer of organic molecules
into the surfactant micelles. When this solution is subjected to
ultrafiltration, the micelles and the associated organic molecules

are retained, while the filtrate contains only very low
concentrations of organic solute and monomer surfactant. A
potential application of this idea is in the removal of toxic
organic materials from waste waters, which would concomitantly
acquire a small load of (presumably biodegradable) surfactant. A
preliminary test of this idea has recently been carried out (4).

It is well known that liquid chromatography separates soluble
analytes by virtue of differences in their distribution between a
stationary phase and a flowing liquid phase. In high performance
liquid chromatography (HPLC), the experimental system is arranged
so as to yield separations with high efficiency and rapid
throughput. In micellar chromatography, also called pseudophase
chromatography (5), the mobile phase contains surfactant micelles.
In this case, the mobile phase, while macroscopically homogeneous,
is microscopically heterogeneous. Analytes now partition between
the microscopically continuous aqueous solvent and the micelles,
in addition to partitioning to the stationary phase. This is
shown schematically in FIG. 1. Micellar chromatography is
commonly employed as a variant of reversed-phase chromatography,

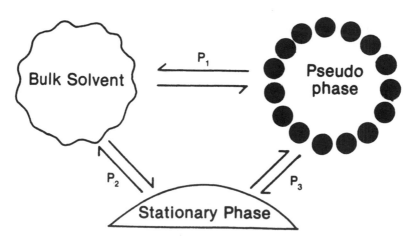

FIG. 1 Representation of the original three-phase model which
allowed a theoretical description of micellar or pseudophase LC.
"P" is the partition coefficient of a solute between the indicated
phases. From ref. (5), with permission.

wherein the surface of the stationary phase is non-polar and
analytes adhering to the stationary phase are progressively eluted
with a mobile aqueous phase containing an organic cosolvent (e.g.,
acetonitrile, n-propanol, etc.). In micellar chromatography, a
micellar system replaces the organic cosolvent. Many useful
separations by conventional reversed-phase and by micellar
chromatography can be carried out isocratically (with constant
solvent composition). In conventional reversed-phase HPLC, the
use of a gradient of solvent composition permits rapid separations
of analytes with a broad range of distribution between the mobile
and stationary phases. In gradient micellar HPLC, the gradient of
organic/aqueous/solvent is replaced by a gradient of surfactant
concentration (6). The surfactant concentration is always
maintained above the CMC, so that there is always a constant
concentration of monomeric surfactant in the mobile phase. The
amount of surfactant bound to the stationary phase does not change
throughout the gradient. This simplifies the theoretical
framework of the chromatography, and also allows a very rapid
re-equilibration of the chromatographic column for repetitive
analyses. An additional advantage of micellar-gradient systems is
that they are compatible with electrochemical detection while
conventional organic solvent gradients are not.

It is claimed, with some justification, that micellar systems
cost less, have lower toxicity, and sometimes offer improved
selectivity over conventional solvent systems. In general,
however, chromatography operates somewhat less efficiently than
conventional reversed-phase chromatography (5).

Attempts to apply micellar chromatography to the separation
of proteins have not been as successful as small molecule
separations (7). The reasons for this have not been fully
established, but it is highly likely that protein denaturation
plays a role. Denaturation implies non-covalent (conformational)
changes in the protein structure, with concomitant loss of
specific biological activity. Denaturation may be time-dependent
(on the time-scale of the separation process), and can sometimes
be reversed by manipulations of the solvent conditions.

The ability of sodium dodecyl sulfate micelles to denature
and solubilize serum proteins has been exploited in the HPLC
analysis of drugs in blood. A number of drugs can be determined
by direct serum injection into a surfactant-containing mobile
phase (8) without prior sample preparation (extraction,
precipitation, etc.). Conventional reversed-phase chromatography
of serum samples leads to precipitation of proteins at the head of
the column or in the column itself, with severe or catastrophic
degradation of the system's performance. Not only does the use of
a surfactant solubilize the proteins, preventing precipitation,
but it also displaces drugs which are often bound to serum
proteins, thereby permitting accurate estimation of the total
amount of each analyte in the sample.

In another context, surfactant denaturation of proteins is
the basis of an electrophoretic method for molecular weight
estimation. Sodium dodecyl sulfate (SDS) forms aggregates with
proteins in which the protein structures are drastically altered
from the native form. As the SDS concentration is increased above
the CMC, the aggregate composition reaches a limiting composition
of 1.4 g SDS/g protein (9). The rate of electrophoretic migration
of this SDS-protein aggregate through a mechanically stabilizing
gel is an inverse function of the logarithm of the molecular
weight of the peptide chain (10). The rates of migration are
largely independent of the amino acid composition of the peptides,
although occasional exceptions have been noted. Under the
conditions of the experiment, the aggregates appear to behave as
if their charge/mass ratios are independent of peptide mass, while
their frictional resistance to migration increases monotonically
with molecular weight.

The separation methods thus far discussed that deal with
biological macromolecules almost without exception lead to
denaturation and loss of biological activity. For many analytical
and for most preparative operations, the objective is to obtain
pure, native, active products. Our laboratory has recently begun
to explore ways in which surfactants at submicellar concentrations
may be used to pursue this objective--the high performance
chromatography of native proteins.

It is generally found that protein solubilizing agents tend
to be denaturing agents. The question is, then, "Is there a
necessary parallel between solubilization potential and
denaturation potential?" We therefore have begun to search for
surfactants and conditions for their use that favor protein
solubilization without denaturation. Scattered empirical evidence
suggests that nonionic and dipolar-ionic surfactants are less
strongly denaturing than are ionic surfactants. Substantial
empirical evidence indicates that surfactant concentrations at or
above the CMC are much more likely to denature proteins than are
submicellar concentrations. It also seems likely that surfactants
with a high CMC are less likely to denature proteins than those
with a low CMC.

We have carried out preliminary HPLC studies using
submicellar concentrations of the surfactant CHAPS (11). This
material, 3-[(3-cholamidopropyl)dimethylammonio]-1-propane
sulfonate, is a dipolar ion derivative of cholic acid, as shown by
its structure in FIG. 2. We chose this material because it has

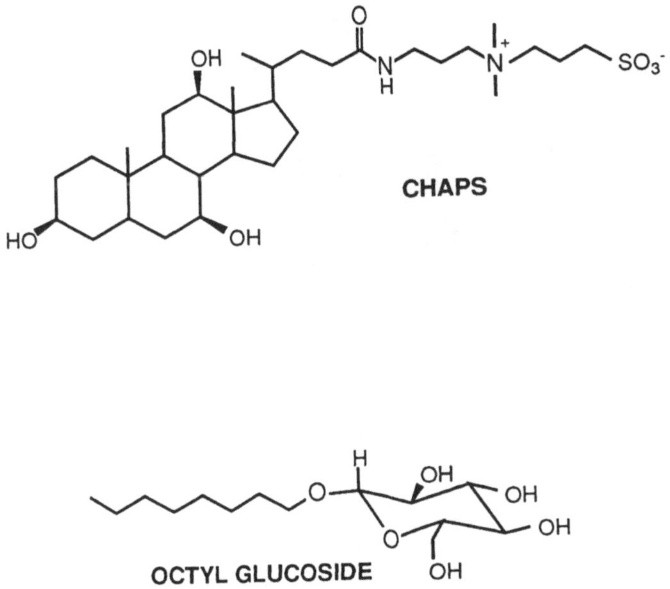

CHAPS

OCTYL GLUCOSIDE

FIG. 2 Two surfactants of zero net charge. Above, 3-[(3-
cholamidopropyl)dimethylammonio]-1-propane sulfonate, CHAPS.
Below, n-octyl β-D-glucopyranoside.

proved to be effective in solubilizing membrane proteins in active
form (12), and because there are additional reasons why this
particular bile salt derivative might meet our requirements.

The molecular structure of CHAPS suggests strong constraints
on the modes of its hydrophobic interaction with proteins. The
rigidity of the steroid ring system, in combination with the
placement of its three hydroxyl groups (FIG. 3), limits the
hydrophobic interactions available to bile salts and the bile salt
derivative CHAPS. These materials form atypical micelles: for
cholate, the tetramer is the first aggregate to form, followed by
higher aggregates as the surfactant concentration is increased
(14). The structural constraints leading to the formation of
atypical micelle systems are, we believe, the same constraints
that limit the hydrophobic interactions with proteins and with
stationary phases. These constrained structures may be contrasted
with the relatively unconstrained structures of surfactants such

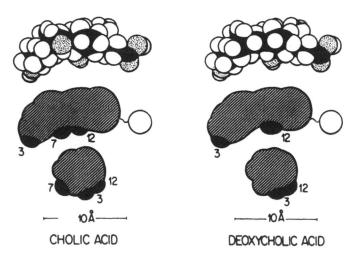

CHOLIC ACID   DEOXYCHOLIC ACID

FIG. 3 Stuart-Briegleb molecular models of the common
unconjugated bile acids with schematic longitudinal and
cross-sectional representations of the models. The stippled atoms
and closed circles and ovals show the positions and orientations
of the hydroxyl functions. From Armstrong and Carey (13), with
permission.

as the n-alkyl glycosides (FIG. 2), sulfates, and ammonium salts, wherein the alkyl groups can display great flexibility.

We carried out isocratic reversed-phase chromatography of three proteins, with several differing concentrations of CHAPS, all below the CMC. The results are shown in FIG. 4, where, in

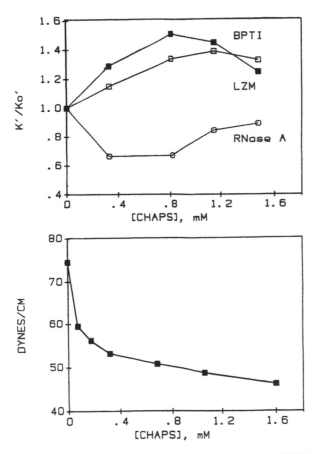

FIG. 4 (upper) Effect of the surfactant CHAPS on retention of three proteins: ribonuclease A, hen egg white lysozyme, and bovine pancreatic trypsin inhibitor. Hydrophobic interaction chromatography was carried out with 1.42 M $(NH_4)_2SO_4$, at $30.0° \pm 0.2°$. The parameter k'/k_0' is the ratio of the capacity factors determined in the presence (k') and absence (k_0') of CHAPS. FIG. 4 (lower) shows the measured surface tension of the several solvents used for the retention experiments of FIG. 4 (upper). The CMC for this system is A ~1.7 mM CHAPS. From ref. (11), with permission.

the upper panel, the normalized retention parameter, k'/ko' is
plotted as a function of CHAPS concentration. (The so-called
capacity factor k', is really a retention parameter:
k'= $(V_r-V_0)/V_0$, where V_r is the retention volume and V_0 is the
void volume of the column. The normalized parameter is defined in
the legend for FIG. 4). From FIG. 4 it is evident that the
surfactant perturbs the retention of the proteins, and that it
shows some selectivity in doing so: two of the proteins are
retarded, but to different extents, while the third is accelerated
in its elution from the column. These experiments have been
repeated with a different kind of chromatographic stationary
phase--in this case elution was hastened for all three of these
proteins. These results serve to emphasize the important role
played by the stationary phase surface. The three proteins used
in these experiments did not undergo any apparent loss of
biological activity in these submicellar surfactant solutions.
However, since these proteins were chosen, among other reasons,
for their high stability, we do not have a fair test of the
denaturation challenge presented by these particular surfactant
solutions.

In the analogous reversed-phase chromatography of small
molecules, a general correlation is seen between a decrease in
surface tension and eluent strength. This has been interpreted in
terms of "solvophobic theory" (15-17), and alternatively in terms
of competitive sorption of the surface-active component of the
mobile phase (18) on the stationary phase. Surface tension is
considered a useful parameter in both formulations of the problem.
On reflection it seems more sensible to try to correlate
interfacial tension (rather than surface tension) with solvent
strength, but suitable interfacial tension measurements for such
systems are difficult to obtain. Comparison of the upper and
lower panels of FIG. 4 shows no obvious correlation between the
surface tension and the retention of any of the three proteins.
We suspect that the same conclusion would be reached if
measurements of the interfacial tension between the mobile and

stationary phase were to be substituted for the present surface
tension measurements. The probable reason for this is that, at
the molecular level, the surface of a globular protein is a mosaic
with non-polar, uncharged polar, and positively and negatively
charged moieties making up the surface. We believe it is unduly
optimistic to expect a macroscopic property such as the surface or
interfacial tension to correlate precisely with interaction
processes involving surfaces of such complexity. On the contrary,
the molecular structures of the interacting surfaces can be
expected to play a strong role in the interaction process. We
would predict, therefore, that n-octyl glucoside and CHAPS at the
same surface (or interfacial) tension will not show the same
chromatographic eluting ability, and that different selectivities
will be seen with each of these structurally different
surfactants.

The questions just raised are of course amenable to
experimental resolution. We have clearly only just begun to
explore several questions that relate to the potential usefulness
of submicellar surfactants in protein separations.

We have drawn FIG. 5 to represent the principal processes in
surfactant-mediated chromatography. The situation is formally
analogous to proposals for ion-pairing mechanisms in the
chromatography of small molecules. However, some additional
complexities should be noted. Even a small protein has the
potential for associating with several ligands--in this case
surfactant molecules. This means that formation of $Pr.Srf_m$ is
more accurately described by a series of equilibria, involving the
successive binding of 1,2,...m surfactant molecules. Second,
surfactant ligands will often form micelles. Third, the protein
can undergo a profound structural disorganization, concomitant
with the binding of n additional surfactant molecules. This
structural disorganization is known as denaturation. Values of
(m+n) can be large, approaching 1/2 the number of amino acid
residues in a polypeptide chain with the surfactant sodium
dodecylsulfate. Fourth, in contrast with the general experience

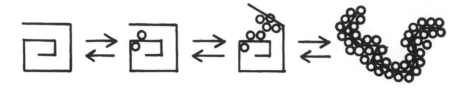

FIG. 5 Scheme for displaying some of the equilibria in
surfactant-mediated protein chromatography. Species bound to the
stationary phase are marked with double-line sub- or superscripts,
e.g., Pr, Srf. Pr: Native Protein, Den.Pr: denatured protein,
Srf: surfactant monomer, Srf_{mic}: micellar aggregate of
surfactant. For simplicity, the drawing separates the lower set
of surfactant interactions from those above the segment of
stationary phase. No such physical separation is implied. From
ref. (11), with permission.

in the chromatography of small molecules, some of the equilibria

in FIG. 5 may be sluggish. If the relaxation time for an

equilibrium step is of the same order of magnitude as the

residence time of the relevant species in the column,

band-broadening or even substantial separation of

slowly-equilibrating species may occur. For the sake of

simplicity we have not included in FIG. 5, the equilibria linked with H^+, nor the fact that the equilibria will depend to varying degrees on the concentration and type of other ions (salt effects). There is no doubt that many of the equilibria and kinetics in question will be responsive to changes in pH and salt. Likewise, a change in temperature can be expected to perturb some of these equilibria. This discussion of FIG. 5 is meant to provide a qualitative description of the interactions of a protein and a surfactant in a chromatographic system, so that the appropriate variables may be recognized and brought under control. In no way do we suggest that the equilibria and kinetic processes need to be dissected out and studied separately to construct a useful separations system.

In this brief discussion, we have outlined a few examples in which the characteristic properties of surfactants can be exploited for separations. Rather general micelle behavior is involved in micelle-enhanced ultrafiltration, in micellar chromatography, and in the sodium dodecyl sulfate gel electrophoresis of proteins. In submicellar surfactant mediated protein chromatography, on the other hand, one seeks particular surfactants whose molecular structures promote protein solubilization without denaturation. The analytical applications of surfactants has a relatively brief history, and the applications have been rather specialized. We are now at a point when the potential of such systems is beginning to be realized, both in general and particular applications.

REFERENCES

1. L. J. Cline Love, J. G. Habarta, & J. G. Dorsey, The Micelle-Analytical Chemistry Interface, Anal. Chem. 56, 1132A-1148A (1984).

2. G. L. McIntire, Electrochemistry in Micelles, Amer. Laboratory 18, 173-180 (1986).

3. K. Ostermaier and B. Dobias, The Separation of Proteins from their Mixtures Using Flotation, Colloids and Surfaces 14, 199-208 (1985).

4. R. O. Dunn, Jr., J. F. Scamehorn, and S. D. Christian, Use of
    Micellar-Enhanced Ultrafiltration to Remove Dissolved
    Organics from Aqueous Streams, Separation Science and
    Technology 20, 257-284 (1985).

5. D. W. Armstrong, Micelles in Separations: A Practical and
    Theoretical Review, Separation and Purification Methods
    14(2), 213-304 (1985).

6. J. G. Dorsey, M. G. Khaledi, J. S. Landy & J. L. Lin,
    Gradient Elution Micellar Liquid Chromatography, J.
    Chromatogr. 316, 183-191 (1984).

7. R. A. Barford & B. J. Sliwinski, Micellar Chromatography of
    Proteins, Anal. Chem., 56, 1554-1556 (1984).

8. F. J. DeLuccia, M. Arunyart & L. J. Cline Love, Direct Serum
    Injection with Micellar Chromatography for Therapeutic Drug
    Monitoring, Anal. Chem. 57, 1564-1568 (1985).

9. A. Helenius, D. R. McCaslin, E. Fries, & C. Tanford, Methods
    in Enzymology Vol. LVI, Acad. Press, N.Y., pp. 734-749
    (1979).

10. K. Weber and M. Osborn, Proteins and Sodium Dodecyl Sulfate:
    Molecular Weight Determination on Polyacrylamide Gels and
    Related Procedures, in "The Proteins" 3rd ed'n, Vol. 1 (H.
    Neurath and R. L. Hill, eds.), Acad. Press, N.Y., 1975, pp.
    179-223 (1975).

11. Wetlaufer, D. B., & Koenigbauer, M. R., Surfactant-Mediated
    Protein Hydrophobic Interaction Chromatography, J.
    Chromatogr. 359, 55-60 (1986).

12. L. M. Hjelmeland & A. Chrambach, in J. C. Benter & L. C.
    Harrison, eds., "Membranes, Detergents, and Receptor
    Solubilization", Alan R. Liss, Inc., N.Y. pp. 35-46 (1984).

13. M. J. Armstrong & M. C. Carey, The hydrophobic-hydrophilic
    balance of bile salts, J. Lipid Res. 23, 70-80 (1982).

14. P. C. Mukerjee, Y. Moroi, M. Murata, & A. Y. S. Yang, Bile
    Salts as Atypical Surfactants and Solubilizers, Hepatology 4,
    No. 5, 615-655 (1984).

15. O. Sinanoglu, Molecular Interactions with Liquids, the
    Solvophobic Force, and Molecular Surface Areas, in
    H. Ratajczak and W. J. Orville-Thomas, eds., "Molecular
    Interactions," Vol. 3, pp. 283-342, Wiley, NY (1983).

16. W. R. Melander, D. Corradini, and Cs. Horvath, Salt-Mediated
    Retention of Proteins in Hydrophobic-Interaction

Chromatography. Application of Solvophobic Theory,
J. Chromatography 317, 67–85 (1984).

17. M. T. W. Hearn and B. Grego, High-Performance Liquid
Chromatography of Amino Acids, Peptides, and Proteins, XLVI.
Selectivity Effects of Peptidic Positional Isomers and
Oligomers Separated by RPHPLC. J. Chromatogr. 266,
75–87 (1983).

18. M. Tang and S. N. Deming, Interfacial Tension Effects of
Non-ionic Surfactants in Reversed-Phase Liquid
Chromatography, Anal. Chem. 55, 425–428 (1983).

Novel Separation Techniques
Group Discussion

John F. Scamehorn, University of Oklahoma, rapporteur

An overview of novel separation techniques utilizing surfactants was given. This overview is expanded in another chapter
in this book.

A great deal of the discussion centered on micellar chròma-
tography, particularly for application involving protein separation. This is currently a laboratory technique and no large
scale processes were known to the group.

Micellar Chromatography: Micellar chromatography is similar to
ordinary reverse phase chromatography. Similar packings and flow
rates are utilized. However, a micellar solution is used as the
mobile phase instead of an aqueous organic mixture. This technique is discussed in detail in another chapter in this book.

One aspect of micellar chromatography with surprising advantages is the use of gradient programming. In this case, the
gradient is the concentration of surfactant in the mobile phase
(or concentration of micelles in the mobile phase). Large
changes in the capacity factor can be attained using gradient
programming. Another advantage of gradient programming is that
there is no significant effect of micellar concentration on the
signal of an electrochemical detector. Therefore, unlike solvent

programming in HPLC, this micellar gradient programming does not
cause drift in the electrochemical detector.

If the surfactant can have specificity built into it, it
could be designed to solubilize some components more effectively
than others. For example, the nonpolar part of the surfactant
could be changed from a flexible tail to something with less
flexibility. Optically active surfactants would also have potential for separating optically active solutes. A naturally
occurring surfactant of this sort might be inexpensive.

Protein Denaturation: Many surfactant-based separations utilize
the solubilization or dissolution of a solute in the intertwined
hydrocarbon tails of a surfactant aggregate (a hydrophobic environment). Proteins tend to denature in organic solutes. Therefore,
when proteins are being purified in these surfactant-based separations, denaturation of these proteins is of concern.

An example of this is micellar chromatography. There is a
high risk of losing enzyme activity or any biological activity
in the presence of micelles. Two conditions which stabilize proteins are low temperatures and high salt concentrations. The salt
must be chosen to have the correct degree of lyotropicity, such
as ammonium sulfate. The presence of stabilizing salts can lead
to problems in using proteins in standard reverse phase HPLC. The
adsorption of the protein on the packing is too high, leading to
long retention times. A possible solution to this is hydrophobic
interaction chromatography, where the packing has less dense
hydrophobic coverage, and to use perhaps a C_3 or C_4 alkyl chain,
rather than the C_{18} chain typical of reverse phase chromatography.

Another separation technique in which denaturation could be a
serious problem is in liquid-liquid extractions using surfactants.
In this technique, conditions (e.g., temperature, additives) are
adjusted so that the solution separates into two phases, one
surfactant-rich and the other surfactant-poor. A solute will tend
to concentrate in the surfactant-rich phase. If the solute is a
protein, the same denaturation whi ch occurs in micelles could
occur, since this phase is really just a concentrated micellar
solution.

One technique which shows promise for separating proteins
without denaturation is the use of inverted or reverse micelles.
These micelles are formed in an organic solvent with the hydrophilic groups in the interior of the micelle and the hydrophobic
groups at the surface of the micelle. A droplet of water can form
at the interior of the micelle surrounded by surfactant hydrophilic groups. While surfactant trapped in this water should not
denature, the limited water solubility of some proteins may limit
the potential of this method.

Adsorption-Based Separations: There are several surfactant-based
separations based on adsorption (see the separations overview
chapter). A major advantage of solution-based separations over
these surface-based separations is that the diffusional effects are
much less important in the former. In the porous solids necessary
to get high surface area, pore diffusion effects can limit the rate
at which a separation process can occur.

8
An Overview of Surfactant-Based Separation Processes

John F. Scamehorn and Jeffrey H. Harwell*

*Institute for Applied Surfactant Research, and
School of Chemical Engineering and Materials Science
University of Oklahoma
Norman, Oklahoma*

I. INTRODUCTION

Surfactant-Based separation processes are a major emerging technological area in both surfactant science and separations science. Surfactant-based separations have a number of potential advantages over traditional methods. They often are low-energy processes because large temperature or endothermic phase changes are not being used to effect separations. Surfactants are often environmentally innocuous and of low toxicity, so that the leakage of a small concentration of surfactant into an aqueous process stream from the separation may be tolerable, in contrast to toxic solvents from liquid-liquid extraction, for example.

* Financial support for the writing of this chapter was provided by Office of Basic Energy Sciences of the Department of Energy grant no. DE-AS05-84ER13175, National Science Foundation grant no. CPE-8318864, the Oklahoma Mining and Minerals Resources Research Institute, and the OU Energy Resources Institute.

Applications   of   surfactant-based   separations
include bioseparations,   where easily degraded materials
must be treated mildly.   Pollution control of water and
air streams can also be done effectively with surfactant
technology.

The goal of this brief overview is to introduce
some of the surprisingly large number of surfactant-
based separations which are known today.   The techniques
are divided into commercially widely used methods (1
method),   methods   with   some   current   commercial
applications (2 methods), and promising techniques which
have yet to reach large-scale commercial application (8
methods).   Our categorization is arguable, but it is
meant to give the chapter structure.

We cover the essentials of each technique and also
give references for readers interested in more details.

II.   PROCESSES   WITH FULL-SCALE   COMMERCIAL UTILIZATION

A. Froth Flotation

From an economic point of view, ore flotation is
currently far and away the most important surfactant-
based separation process,   so much so that froth
flotation is frequently used as a synonym for ore
flotation. This is not surprising in that an estimated
billion tons of ore is processed by this method each
year.

The basic principle in froth flotation is quite
simple: bubbles of air sparged into a column of water
rise to the to the top of the column due to the bouyance
force. Surfactants adsorbed on the surface of small
particles such as ore fines cause the particles to
adhere to the surface of the rising bubbles. The
particles are then floated to the top of the column

where they collect in the froth generated by the sparged air. The froth can then simply be skimmed off to complete the removal of the particulate matter from the process stream. If two or more different kinds of particles are present in the process stream, then they can be separated from one another if the surfactant preferentially adsorbs on one kind of particle over the other.

In practice it may be necessary to add "activators" or "depressers"--usually either potential determining ions of the solids or divalent cations to increase anionic surfactant adsorption--in order to obtain the proper extent of surfactant adsorption necessary to produce good flotation. "Foamers", such as a long chain normal alcohol, may also be added to stabilize the foam.

In addition to a large body of excellent scientific literature, a great deal of art has accumulated concerning the application of this technique to ore processing, and the interested reader should consult a standard reference [1-3].

III. PROCESSES WITH SOME COMMERCIAL APPLICATIONS

A. Absorptive Bubble Separations

The same basic idea used in ore floatation can also be used advantageously for a wide variety of other separations. In each case the goal of the process is to cause the material being removed from solution to adhere to the surface of bubbles rising through the solution. Separations of this type can be conveniently divided into particulate flotations and colligend flotations. Particulate flotations deal with undissolved material, colligend flotations with dissolved material. Both types of materials can be efficiently removed by flotation.

Ore flotation can actually be thought of as a specific example of particulate flotation. Other examples of successful application include removal of bacterial spores, algae, and precipitates. In each case, successful flotation requires selection of a surfactant that will adsorb at the particle/solution interface, which implies proper selection of the charge of the surfactant and either adjustment of the pH or addition of an ion to promote adsorption of the surfactant on the particle surface.

In colligend flotation, the surfactant is chosen so that its adsorption at the air/water interface creates an environment that is favorable to the adsorption of the material to be removed from solution. For example, ppb concentrations of metal ions may be removed from solution by adding anionic surfactant to the solution to be foamed; adsorption of the surfactant at the air/water interface causes the metal ions to be entrained in the electrical double layer of the rising bubble, and thus floated from solution. Similarly, organic acids or other organic compounds may be incorporated into the surfactant layer by adsolubilization. We may also class as colligend flotation successful demonstration of the concentration of albumin [4], separation of diatase from liase, and of urease from catalase.

Though most well known as a batch process applied on a laboratory scale, industrial scale continuous flow units have been developed, and design procedures for these units have appeared in the literature [5-6]. This process is likely to become increasingly important over the next few decades.

B. Surfactant-Based Liquid Membranes

In the preparation of liquid surfactant membranes [7-9], an emulsion is formed between two immiscible phases. The emulsion is then dispersed as droplets into a third phase which is immiscible with the external phase of the emulsion. The component which needs to be removed from the continuous phase can then diffuse through the external phase of the emulsion globule to the encapsulated droplet or the internal reagent phase where the component is trapped or converted by chemical reaction.

For example [7], to remove ammonia from water, an oil continuous emulsion is formed with sulfuric acid as the emulsified or internal reagent phase. These emulsion globules are then dispersed in the water. The ammonia diffuses through the oil phase to the sulfuric acid droplets, where an oil-insoluble ammonium ion is formed and left trapped in the sulfuric acid droplet.

In liquid surfactant membranes, the external phase in the emulsion (e.g., the oil in the above example) is considered the membrane. Surfactant is added in the emulsion to control the stability, permeability, and selectivity of the membrane [7].

When ions are being removed from water, the emulsion reagent phase may consist of some other ion of similar charge. A Donnan equilibrium is established across the membrane phase using an ion exchange reagent, resulting in a concentration of the ion of interest in the reagent phase. For example, extraction of the cupric ion from water may be balanced by the counter transport of protons [9].

At some point, the emulsion must be separated from the solution and the reaction product or extracted component removed before the emulsion can be recycled back to the process.

Liquid surfactant membranes have great potential in such applications as removal of trace organics or heavy metals from water, fractionation of hydrocarbons, biochemical separations, and as membrane reactors [9].

IV. LABORATORY PROCESSES WITH HIGH POTENTIAL

A. Absorption into Emulsion Solutions

In an automobile spray painting process, fugitive
hydrocarbon solvents must be removed from air before its
release to meet emission control requirements. In a new
process to achieve this, the air is bubbled through an
oil-in-water emulsion [10]. Surfactants are used to
stabilize the emulsion. The oil-soluble solvents are
dissolved in the oil droplets while the water soluble
components are dissolved in the continuous phase of the
emulsion. A small pH adjustment then causes the
emulsion to break, forming an oil phase, a water phase,
and a sludge. The solvents being removed from the air
are concentrated in the sludge. Residual solvents in
the oil phase can be recoved by thermal methods. Once
the fugitive solvents are removed as sludge or from the
liquid phases, the emulsion can be reformed by
readjusting the pH, for recycle to absorb more solvent.
Many emulsion forming/breaking steps seem possible
without degradation of the oil or surfactant.

This process shows great potential for emmission
control in auto spray painting, but numerous other
applications are possible, since removal of organics
from air is a commonly encountered problem.

B. Micellar-Enhanced Ultrafiltration

Dissolved organics and multivalent ions (e.g.,
heavy metals) must frequently be removed from water in
industrial operations. Micellar-enhanced
ultrafiltration (MEUF^{TM}) is a recently developed
membrane separation which has been shown to effectively
remove slightly soluble organics and multivalent heavy
metals from water [11-14].

In MEUF^{TM}, surfactant is added to the water stream.
The surfactant forms micelles, spheroidal aggregated

structures containing about 80-100 surfactant   molecules
for the surfactant types of interest. The micelle has a
hydrocarbon-like   interior,   with   the   surfactant
hydrophilic groups on the surface of the micelle exposed
to the aqueous solution. Any dissolved organics in the
water will   tend to   dissolve or   solubilize in   the
hydrophobic interior of the micelles. If the surfactant
is anionic, the micelles have a large negative charge.
Any cationic multivalent metal ions in solution will
tend to bind to the surface of these micelles.

The solution is then treated in an ultrafiltration
unit using membranes with pore sizes just small enough
to reject the micelles. The solubilized organic solute
and/or the bound metal ion associated with the micelle
are also rejected by the membrane.   The   permeate
solution which passes through the membrane can be very
pure. For example, tert-butyl phenol concentration [1-
3] in the permeate has been shown to be reduced by a
factor of 100 compared to the   feed   solution. Divalent
copper [14] or zinc [13] concentrations have also been
shown to be reduced by a factor of 100 in one pass.

The retentate solution contains   the   rejected
organic solute, metal ions, and surfactant in high
concentrations. Relative water   flux   rates   in MEUF^{TM}
have been shown to be high, with the membrane fluid
dynamics exhibiting classical concentration polarization
behavior with a well-defined gel layer [12].

MEUF^{TM} shows great promise for removal of toxic
heavy metals and organics from wastewater simultaneously
and for removal of valuable biochemicals produced in
dilute concentrations in bioreactors.

C. Extraction into Reverse Micelles

It has been demonstrated that a number of proteins
can be solubilized in reverse micelles, then recovered
without irreversible denaturation or loss of activity.

Recently, it has been recognized that this phenomenon
has great promise as the basis for a new, continuous
flow bioseparation technique [15].

The potential of the technique can be recognized by
noting its similarity to solvent extraction, one of the
most important industrial separation techniques. While
solvent extraction is not direcly applicable to
separations of proteins, because most proteins are
insoluble in apolar solvents, the same proteins can be
readily solubilized in reverse micelles.

In the first stage of the process, the protein is
transferred into reverse micelles from the original
aqueous process stream by contacting with the apolar
solvent in the presence of surfactant. Solubilization
appears to be governed by minimization of the
electrostatic contribution to the free energy of the
system, so that low electrolyte concentrations and pH
values at which the protein is charged tend to favor
transfer of the protein into the reverse micelle. After
the two phases have been separated, the organic, apolar
solvent phase, now containing the protein solubilized in
the reverse micelles, can be contacted with another,
lower volume aqueous stream at values of pH and
electrolyte concentration which favor transfer of the
protein back into aqueous. The protein thus may be
separated and concentrated in the same process.

D. Coacervate-Based Separations

Coacervate-based separations are useful in the
removal of dissolved organics from water [16-18]. the
basis of this process is a well known surfactant
phenomenon, the separation of aqueous solutions of
nonionic surfactant into a concentrated phase containing
most of the surfactant (coacervate phase), and a dilute
aqueous phase containing low concentration of

surfactant. This separation spontaneously occurs at temperatures above the "cloud point" of the surfactant at the given conditions.

In application, surfactant is added to the aqueous solution which contains the dissolved organic. The surfactant is chosen so that the solution temperature is above its cloud point. When the solution separates into the coacervate phase and the dilute phase, the dissolved organic will tend to be much more soluble in the coacervate phase. Conditions can be arranged so that the coacervate phase is an isotropic solution, in which case it is similar to a concentrated micellar solution [17]. Alternatively, the coacervate may be a liquid crystalline phase [16]. This process is simply a liquid/liquid extraction of the organic between the dilute phase and the coacervate. The organic can be substantially concentrated into the coacervate liquid compared to its concentration in the original water stream. Using a stagewise operation, a very high degree of separation can be achieved. This process has applications in pollution control [17] and biotechnology [16].

E. Micellar Chromatography

In high performance liquid chromatography (HPLC), selectivity may be affected by the affinity of the various solutes being separated for the stationary and for the mobile phases. In micellar chromatography, the mobile phase contains surfactant in micellar form [18,19]. The tendency of the various solutes to solubilize or dissolve in the micelles may differ substantially, resulting in different retention times in a chromatographic column. For example, in the process of solubilization of solute in a micelle composed of ionic surfactant, both hydrophobic and electrostatic forces may be important, resulting in a potentially large effect of solute and surfactant structure on solubilization.

Examples of solutes which have been separated using micellar chromatography include pesticides, amino acids, nucleosides, phenolics, polycyclic aromatic hydrocarbons, and fatty acids [18].

It seems feasible that large scale preparative micellar chromatography may evolve with the emerging need to separate valuable biotechnology products.

F. Admicellar Chromatography

In admicellar chromatography, one again utilizes the preferential partitioning of one or more components in a dilute aqueous solution into a surfactant aggregate in order to effect a separation. Whereas in micellar chromatography the surfactant aggregate is a mobile micelle, in admicellar chromatography the aggregate is an immobile admicelle. By manipulation of parameters like surfactant charge and structure, and concentration of potential determining ions for the solid phase, a substantial fraction of the solid surface can be covered with adsorbed surfactant aggregates, called admicelles for their micellar-like properties.

By co-introducing into a fixed bed unit surfactant and a target compound, such as a pollutant or protein, the target compound will be adsolubilized in the immobile admicelles. If several organic compounds are present, there will also be a preferential partitioning of the compounds between the aqueous phase and the admicellar phase [20].

As the bed nears saturation with the organic, the admicelle layer, along with any adsolubilizate, may be desorbed by changing the concentration of potential determining ions in the inlet stream. Depending on the chromatographic velocity of the desorption wave, concentrations of the target compound in the effluent may even exceed the solubility limit of the compound in water, resulting in a spontaneous phase separation [20].

This process can be adapted to gas phase separations by first depositing the surfactant layer by adsorption from solution under admicelle forming conditions. Then the aqueous phase is displaced by either air or by the process stream itself. Organic components of the process stream will then preferentially partition into the thin adhering layer of surfactant. When the bed becomes saturated, the aqueous stream is re-introduced, but now at conditions favoring desortpion of the admicelle layer.

At the very least, this process offers an attractive alternative to removal of pollutants by fixed-bed carbon adsorption, since the bed is readily regenerated **in situ** by a low energy process. In many cases, however, it will also make possible the recovery of the pollutant at the high effluent concentrations which are obtained, so that a waste treatment process actually produces a marketable product. In bioseparations it has the attractiveness of potentially combining a separation step with a concentration step in a single unit.

G.   Polymerized Thin-Film Separations

In admicellar chromatography the surfactant layer on the surface of the inorganic packing can be easily removed. Another application of the phenomenon of adsolubilization of compounds in admicelles is the use of monomer as adsolubilizate, followed by in situ polymerization of the monomer inside the admicellar layer. This results in formation of an insoluble, ultra-thin organic layer on the surface of an inorganic support [21-23].

Advantages of this new process over conventional processes are numerous. Organic packings tend to crush or compress in industrial scale applications, which contributes to large pressure drops in fixed-bed applications, and thereby to high energy consumption through pumping costs. This is a major motivation for the use of fluidized beds in, for example, ion exchange. Inorganic packings, when manufactured in such a way as to increase the available surface area per unit weight, avoid this complication. Whereas the inorganic packings in and of themselves will generally not exhibit sufficient adsortive capacity from aqueous solutions to be useful for separations, they can be used as supports for organic films.

Films formed by polymerization in admicelles appear particularly promising for adaptation of inorganic packings to, for example, bioseparations. The films can be formed from any monomer which can be solubilized. They can also be formed inside any pores of the solid into which surfactants will diffuse, so that high surface area packings are readily formed by this process. Perhaps most important, however, is that because the admicelles are self-assembling, we expect a high degree of molecular organization within the layer. The possibility of forming an organized molecular layer on an inorganic support promises the ability to achieve remarkably high degrees of selectivity in a separation, including the possibility of industrial scale resolution of stereoisomers by diasteriameric associations in a film formed from optically active surfactants.

H. Surfactant-Enhanced Carbon Regeneration

Adsorption beds containing activated carbon are widely used in industry to remove organics from water and air. One of the major disadvantages of this method

is the difficulty of regeneration of the carbon once
saturated. When the organic is highly volatile, in situ
thermal regeneration (e.g., hot steam) can be used.
However, often the carbon must be removed from the bed
and the organic burned off in regeneration furnaces. A
universal in situ regeneration process would be
valuable.

One such method is surfactant-enhanced carbon
regeneration (SECR) [13,24]. In SECR, a concentrated
surfactant solution is passed through the saturated
carbon bed. The adsorbed organic solute desorbs and is
solubilized in the micelles formed by the surfactant.
The concentrated surfactant solution may have a large
solubilization capacity for the organic. Therefore, a
much smaller volume of surfactant solution can be used
than the volume of the water originally treated for
aqueous phase applications. After the organic is
removed from the carbon, water is used to rinse residual
surfactant from the carbon. An advantage of SECR over
solvent regeneration is the relatively innocuous nature
of surfactants from an environmental point of view.
Thus, the product of the rinse step can be directed to
the normal sewage treatment system, since it contains
only surfactant. The regenerated carbon would then be
ready for reuse for liquid phase operations. If the
carbon is being used for gas phase applications, the
carbon can be dried after the rinse step before reuse.

In an example of a application of SECR, a
surfactant solution equal to 0.05 % of the volume of the
original water treated by the bed was shown to be
capable of removing 70 % of the organic (tert-butyl
phenol) which was on the carbon when saturated. The
regenerant solution therefore had a factor of 1000
higher concentration of the tert-butyl phenol than that
in the original water treated. These results

demonstrate that the technical concept of SECR is valid; spent carbon can be regenerated by concentrated surfactant solutions.

V. SUMMARY

The separation processes outlined here utilize a wide range of phenomena involving surfactants to effect separations. All but two concern removing components from a liquid phase; absorption into emulsion solutions involves removal of organics from a gas phase, and admicellar chromatography may be adapted to gas phase separations.

Another way to characterize these techniques is by the surfactant aggregate structure effecting the separation:

Micelles: micellar-enhanced ultrafiltration; extraction into reverse micelles; micellar chromatography; and surfactant-enhanced carbon regeneration.

Adsorbed surfactant aggregates: froth flotation; admicellar chromatography; chromatography based on polymerized surfactant bilayers.

Emulsions: liquid surfactant membranes; absorption into emulsion solutions.

Foams: absorptive bubble separations.

Coacervate: coacervate-based separations (may be either micellar solutions or liquid crystals).

This class of separations has so much potential in both classical applications and emerging technology that surfactant-based separations will likely become a major area of technological development in chemical engineering in the next several decades.

An exciting aspect of the innovation which will be seen during this period is the coupling of these techniques to affect a final separation. To select a single example, while a foam separation might look promising from the point of view of selectivity or as a final purification step, the volume of solution which might need to be treated might make capital expenditures prohibitive. Application of micellar enhanced ultrafiltration or admicellar chromatography to the process stream could then be used to increase the concentration of target compound in the inlet stream by two or more orders of magnitude. This would make possible a significant down-scaling of the foam separation unit, with a concomitant reduction of capital costs. We have not even attempted an overview of such synergistic possibilities here.

REFERENCES

1. M.C. Fuerstenau (ed.), **Flotation. A.M. Gaudin Memorial Volume.** Am. Inst. Min. Met. Petrol. Eng., New York, N.Y., (1976).

2. B.R. Palmer, B.G. Gutierrez, and M.C. Fuerstenau, **AIME Trans.**, **258**, 257 (1975).

3. A.R. Laplante, J.M. Toguri, and H.W. Smith, Int. J. Miner. **Process.**, **11**, 203 (1983).

4. S.I. Ahmad, **Sep. Sci.**, **10**, 673 (1975).

5. R. Lemlich (ed.), **Adsorptive Bubble Separation Techniques**, Academic Press, New York, 1972.

6. Y. Okamoto, and E.J. Chou, in Handbook of Separation Techniques for Chemical Engineers, (P.A. Schweitzer, ed.), McGraw-Hill, New York, 1979, Chap. 2.5.

7. T.A. Hatton, E.N. Lightfoot, R.P. Cahn, and N.N. Li, Ind. Eng. Chem. Funda., 22, 27 (1983).

8. N.N. Li, D.T. Wasan, and Z. Gu, in Surfactants and Chemical Engineering (D.T. Wasan, D.O. Shah, and M.E. Ginn, ed.), Marcel Dekker, New York, In Press.

9. W.S. Ho, T.A. Hatton, E.N. Lightfoot, and N.N. Li, AICHE J., 28, 662 (1982).

10. W.H. Lindenberger, in Surfactants in Emerging Technologies (M.J. Rosen, ed.), Marcel Dekker, New York, In Press.

11. R.O. Dunn Jr., J.F. Scamehorn, and S.D. Christian, Sep. Sci. Technol., 20, 257 (1985).

12. R.O. Dunn Jr., J.F. Scamehorn, and S.D. Christian, Sep. Sci. Technol., In Press.

13. J.F. Scamehorn, and J.H. Harwell, in Surfactants and Chemical Engineering (D.T. Wasan, D.O. Shah, and M.E. Ginn, ed.), Marcel Dekker, New York, In Press.

14. J.F. Scamehorn, R.T. Ellington, S.D. Christian, B.W. Penney, R.O. Dunn Jr., and S.N. Bhat, AICHE Symp. Ser., In Press.

15. K.E. Goklen, and T.A. Hatton, Biochem. Biophys. Res., 1, 69 (1985).

16. R. Heusch, Intern. Z. Biotech., 3, 2 (1986).

17. N.D. Guillickson, J.F. Scamehorn, and J.H. Harwell, in **Surfactant-Based Separation Techniques** (J.F. Scamehorn, and J.H. Harwell, ed.), Marcel Dekker, New York, In Press.

18. D.W. Armstrong, **Sep. Purif. Meth.**, 14, 213 (1985).

19. D.B. Wetlaufer, in **Surfactants in Emerging Technologies** (M.J. Rosen, ed.), Marcel Dekker, New York, In Press.

20. T.P. Fitzgerald and J.H. Harwell, **AIChE Symp. Ser.**, In Press.

21. J. Wu, J.H. Harwell, and E.A. O'Rear, submitted to **J. Am. Chem. Soc.**.

22. J. Wu, J.H. Harwell, and E.A. O'Rear, submitted to **J. Phys. Chem.**.

23. J. Wu, J.H. Harwell, and E.A. O'Rear, **Colloids and Surfaces.**, In Press.

24. D.L. Blakeburn, and J.F. Scamehorn, in **Surfactant-Based Separation Techniques** (J.F. Scamehorn, and J.H. Harwell, ed.), Marcel Dekker, New York, In Press.

9
NALCOs Hydrocarbon Emission Control Process
A Novel Pollution Control Method

William H. Lindenberger

NALCO Chemical Company
Naperville, Illinois

INTRODUCTION

Surfactants are used extensively in industrial applications for
product formulation, process control, particle size control, and
surface modification. In the field of pollution control, however,
surfactants are generally the cause of problems because surfactants stabilize oil dispersions or emulsions which cause oily
wastewater pollution. These problems have concerned industry for
years. Emulsions formed using ionic surfactants, such as carboxylates, sulphonates and quaternary amines, are readily handled.
However, emulsions formed with nonionic surfactants, such as
ethoxy- and propoxy alkyl adducts, cause very stable emulsions and
require clever chemical methods to economically address waste
treatment. There are not many good chemical solutions to this
problem available in the marketplace.

This presentation will describe how Nalco has applied emulsion breaking expertise to a process which scrubs volatile organic compounds, VOC, from the exhaust air of an automotive
spray painting process. Nalco's process, the Hydrocarbon Emis-

sions Control Process, provides a cost-effective, viable solution
to a major air pollution problem facing the United States automotive industry today.

GOVERNMENT REGULATIONS: THE CLEAN AIR ACT

In October of 1979 standards of performance were proposed under the
Clean Air Act to limit emissions of volatile organic compounds
from automotive surface coating processes. A final standard of
15.1 lbs. of VOC per gallon of applied coating solids was established by the Environmental Protection Agency in 1980 for existing plants and 12.2 lbs. for any new plants under construction.
The best available technology needed to meet these standards requires the use of water-based paints, a coating in which water is
substituted for most of the organic solvent. The equation used
for calculating compliance with these standards is shown below.

$$\frac{\text{Pounds VOC}}{\text{Emitted}} = \frac{\text{(Pounds VOC per Gallon of Coating)}}{\text{(\% Volume Solids) X (Transfer Efficiency)}} \quad (1)$$

In other words, the solvent emitted by the plant is not related to
the volume of air discharged, but rather to the amount of solvent
in the coating and the efficiency of application, or transfer
efficiency. The limit of 15.1 lbs. of VOC was determined using
2.8 lbs. VOC per gallon, 62% volume solids and 30% transfer efficiency for a water-based coating. Data for other coatings, such as
high solids solvent enamel, dispersion lacquers, or DL, and base
coat/clear coat, are shown in Table I.

THE AUTOMOTIVE PLANT: COATING PROCESS

An integral part of the automotive assembly process is surface
coating, which is used for corrosion protection, sound deadening

TABLE I Data for calculating water-borne equivalency

	Water-Borne Enamel	High Solids Solvent Enamel	17% DL	22% DL	27% DL	Base Coat	Clear Coat
lbs. VOC/gal.	2.8	3.6	5.8	5.4	5.0	4.6	4.2
% Vol. Solids	62	50	17	22	27	32	44
% Transfer Efficiency	30	40-60	35		40-60	40	62
Emission Rate (lbs. VOC)	15.1*	12-18	98		31-46	36	15

*Existing Plant = 15.1 lbs. VOC

New Plant = 12.2 lbs. VOC

and asthetics. Coatings are applied by spray painting, using auto-
matic spray guns, manual spray guns and robotics. In order to provide a safe work environment, a well ventilated paint spray booth
circulates 100 cubic feet per minute per square foot of filtered
air over the worker. In a typical plant this can reach 200,000
cubic feet per minute of exhaust air. The exhaust air containing
volatile organic compounds is the regulated source. A schematic
of an automotive paint spray booth is shown in Figure 1.

THE AUTOMOTIVE INDUSTRY PROBLEM

The domestic automotive manufacturer in 1980 was very concerned
with Japanese and German import competition. These "quality" cars
were perceived by many as being superior to the domestic products
and "depth of finish" of the surface coating was critical to this
perception. Water-based coatings would meet compliance with the

FIG. 1 *Schemetic of a paint spray booth.*

Clean Air Act, but could provide neither the "depth of finish" nor
be applied by conventional application techniques. In order to
compete with these foreign manufacturers, a base coat/clear coat,
BC/CC, finishing process was needed. Unfortunately, this coating
could only achieve an emission rate of 20-25 lbs. VOC per gallon of
solids applied. Abatement, or the add-on capture of VOC in excess
of 15.1 lbs., would be required. The following equation describes
the influence of add-on abatement.

$$\frac{\text{Pounds VOC}}{\text{Emitted}} = \frac{(\text{Pounds VOC per Gallon of Coating}) \times (1 - A)}{(\text{\%Volume Solids}) \times (\text{Transfer Efficiency})} \quad (2)$$

$$\text{where} \quad A = \text{Abatement Efficiency}$$

The automotive industry had only one proven abatement technology,
carbon adsorption, which would cost the industry 4-6 billion
dollars to install. This represented a cost of $20,000 per ton of
VOC abated, or as much as $10 per pound of solvent. Several alter-
native technologies already had been evaluated, including,

   biological oxidation of the recirculating water,

   oil scrubbing of the exhaust air, and

   aqueous surfactant scrubbing of the exhaust air.

None of these methods had proven successful in pilot-scale production testing.

NALCO'S APPROACH TO THE HYDROCARBON EMISSIONS PROBLEM

In early 1980, Nalco assembled a team of researchers including
chemists, engineers and marketing personnel to address the abatement problem faced by the automotive industry. The team was given
the goal of developing a chemical solution using existing plant
equipment that would be cost-effective. Several criteria were established.

   The paint spray booth would be the air scrubber.

   The chemical treatment could not effect the coatings process.

The chemical treatment could not effect the finished paint
job.

Abatement had to be measurable.

Abatement had to achieve Clean Air Act requirements for
BC/CC and high solids enamel coatings.

The method had to be cost-effective.

Several screening methods were used. One method utilizing a gas
washing bottle is illustrated in Figure 2.

This method demonstrated the adsorption characteristics of numerous water additives as a function of single solvent and solvent
mixtures. A sampling of the results are shown in Table II.
Analyses were performed using carbon disulfide desorption from
charcoal tubes followed by gas chromatagraphy. Single solvent
tests made the analytical job relatively uncomplicated.

The gas washing bottle test, however, could not simulate the adsorption efficiency expected when spraying paint coatings; it only
demonstrated the effects of gas scrubbing. A second screening
method was devised, which is illustrated in Figure 3.

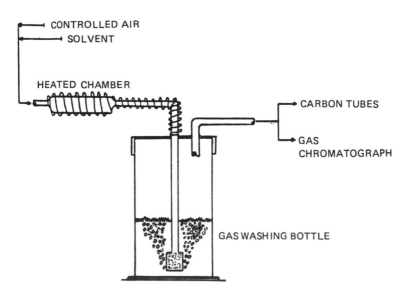

FIG. 2 *Gas washing bottle apparatus.*

TABLE II *Gas washing bottle test results*

Scrubbing Media	Toluene Capture
Water	*5%*
Water/activated carbon	*95%*
Water/clay	*25%*
30% oil-in-water emulsion	*45%*
15% oil-in-water emulsion	*20%*
5% oil-in-water emulsion	*17%*

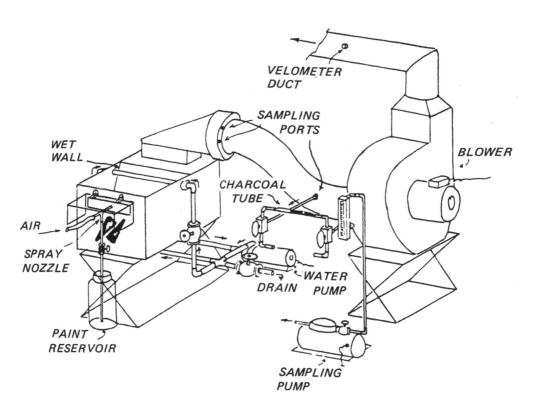

FIG. 3 *Small scale test paint spray booth.*

This screening tool permitted the spraying of water-based coatings, high solids enamels, dispersion lacquers, and the base coat/clear coat coatings. Analyses were much more complex because of the solvent mixtures typically found in commercial automotive coatings. The capture efficiencies were determined by solvent adsorption on charcoal tubes and by direct air sample injection into a Hewlett Packard 5880 gas chromatagraph. Some characteristic results, using an oil-in-water emulsion versus a conventional spray booth water treatment, are illustrated in Table III. This test method shows the oil-in-water emulsion to capture 25% VOC versus 1% for the conventional water treatment program, while generating 30% less sludge.

THE EMULSION APPROACH TO VOC ABATEMENT

After numerous idea generation sessions and the bench scale screening of these ideas, a single technology showed technical

TABLE III *Example of small pilot spray booth test results*

Paint:	*95.0% blaze red*	
	2.5% methylethylketone	
	2.5% toluene	
System:	*Center draft pilot spray booth*	
	250 gm of paint sprayed	
Performance:	*30% oil-in-water emulsion versus conventional*	
	spray booth water treatment	

	O/W Emulsion	Conventional treatment
VOC capture	*25%*	*1%*
Sludge volume	*140 gm*	*200 gm*
Sludge composition		
water	*50%*	*53%*
solids	*20%*	*47%*
oil	*30%*	*0%*

promise. The use of an emulsion, formed in the paint spray booth
water recirculating system, was selected for scale-up. The
process would use the paint spray booth as a wet scrubber, utilize
low volatility oil and thus not affect the coatings process or the
finished job, and mass balance measurement of VOC appeared feasible. Cost effectiveness was highly dependent upon the reuse of
oil from the emulsion.

The emulsion process works by capturing paint solvent from both
the air and the oversprayed paint particles. The emulsion is removed from the recirculating system, resolved into separate oil
and water phases, the VOC is thermally stripped from the oil, and
the oil and water are re-emulsified and returned to the system.
Recovery and reuse of the oil is required and the emulsion's
ability to be formed, resolved and reformed many times makes this
possible. Illustrations of both the batch and continuous processes are shown in Figures 4 and 5, respectively.

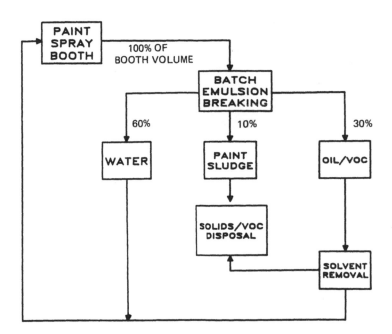

FIG. 4 *Batch emulsion process.*

CONTINUOUS EMULSION PROCESS

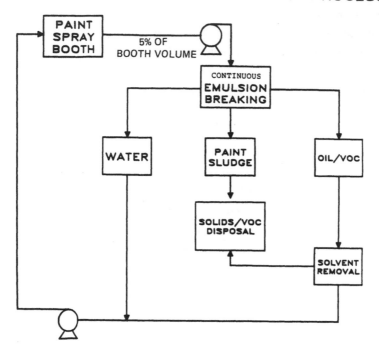

FIG. 5 *Continuous emulsion process.*

TABLE IV *Examples of emulsifiers used in the formation of pH sensitive emulsions*

Oleic acid
Dodecylbenzene sulfonic acid
Petroleum sulfonate
Polyoxyethylated fatty alcohol
Chlorinated stearic acid
Diethoxylated coco fatty amine
Stearic acid
Palmitic acid
Myristic acid
Lauric acid
Tall oil acid
Tallow acid

The selection of surfactants in formulating an emulsion that would
survive repeated "making and breaking" and also be thermally
stable to vacuum stripping was critical. In order to simplify
the process, the emulsion was made pH sensitive. This accommo-
dated the use of sulphuric acid and sodium hydroxide as the only
"triggers" needed to make and break the emulsion. Numerous ionic
emulsifiers were evaluated as illustrated in Table IV.
The oil-in-water emulsifiers selected for this process have HLB
values of 10-30 and form an emulsion containing 5-30% oil upon
shear and pH adjustment. Optimum pH for forming the emulsion is
8.0-9.5 producing a particle size ranging from 2-10 microns. The
emulsifiers can be anionic, cationic and nonionic, but nonionic
emulsifiers form emulsions that are difficult to resolve. Preferred emulsifiers are saturated and unsaturated fatty acids of
16-24 carbon atoms. Sulfonated surfactants are effective, but
require considerably more acid to break the emulsion. In order to
form an effective emulsion 1.5%-10% surfactant is required and pH
must be closely controlled. Selection of the appropriate emulsifier also requires careful attention to the ionic strength of the
aqueous phase. Greater than 500 ppm Ca^{+2} and/or Mg^{+2} or 1.5%
Na_2SO_4 causes emulsion instability.

The emulsion system was tested on a pilot scale using two different paint spray booths, a back section Binks spray booth and a
center draft Flakt spray booth. The results are illustrated in
Table V.
Capture of VOC was dependent on: (1) the concentration of oil in
the emulsion, (2) the level of solvent allowed to accumulate in
the oil phase, (3) the ionic strength of the aqueous phase,
(4) the composition of the solvent in the paint, (5) the type of
paint sprayed, and (6) the transfer efficiency. Oil recovery was
found to be 95%-97% and recovered solvent was clean and potentially reusable. The paint sludge formed in this process was unique.
Instead of a tacky, solid mass, the sludge was a pumpable fluid
with a fuel value of 10,000-15,000 BTU/lb. The estimated cost
for this process, including capital expenditures, is $1,500-2,000
per ton of VOC abated, only 10% of the cost of carbon adsorption.

TABLE V *Performance of the Nalco HEC process*

Paint Type	Specified Transfer Efficiency	Abatement Necessary to Meet Compliance at Specified Transfer Efficiency				Experimental Abatement Results	
		Existing Facility		New Facility		20% Emulsion	30% Emulsion
		Oven Incineration					
		Yes	No	Yes	No		
Lacquer	40%	60%	90%	68%	98%	42%	68%
	60%	38%	52%	51%	81%		
Enamel	40%	0%	22%	13%	44%	51%	64%
	60%	0%	0%	0%	0%		
Base Coat/ Clear Coat	40%	28%	58%	43%	73%	45%[1]	65%[2]
	52%	6%	36%	25%	55%		
	60%	0%	21%	13%	43%		

1 *Using Methyl amyl ketone solvent*

2 *Estimated*

MECHANISM OF THE EMULSION PROCESS

The emulsion process acts by absorbing volatile organic compounds, paint resins, and pigments into the oil phase of the water continuous emulsion. Initially the paint overspray particle contacts the dispersed oil droplet and adheres to the surface. Because the paint particle is essentially organic, it migrates into the oil phase. This explains the observation that the paint sludge appears to "disappear". The VOC, paint resin and pigments distribute themselves based upon their relative solubility within and on the surface of the oil and in the water. Some of the VOC is water soluble and migrates into the water phase. Both the resin and pigments can have hydrophilic and hydrophobic character and thus migrate into the water, remain in the oil, or locate at the oil/water interface. This is illustrated in Figure 6.

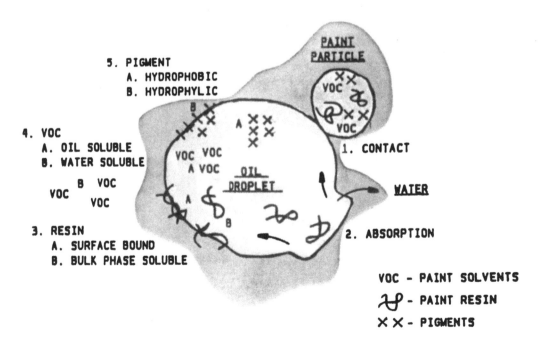

FIG. 6 *Mechanism: HEC emulsion process.*

When the emulsion is broken with acid, three distinct phases are
formed. The oil phase contains most of the VOC captures. A
denser sludge phase contains resin, pigments, VOC, oil and water.
The third phase is water and will contain some VOC and oil. Many
operating mechanical parameters control maximizing VOC recovery
and minimizing oil loss.

CONCLUSIONS

The Nalco Hydrocarbon Emissions Control process is a viable, cost-
effective method of capturing fugitive solvents from an automotive spray painting process. The process utilizes emulsion chemistry and the selection of specific surfactants which must be
carefully chosen in order to perform under rigorous process conditions. The process yields a recovered, valuable solvent, a
sludge by-product that can be efficiently incinerated, and mini-
mizes the need for major capital expenditure. The emulsion used
in this process and the process itself are covered under three
United States patents. [1, 2, 3]

REFERENCES

1. W. H. Lindenberger, et al., *U.S. Patent 4,378,235* (1983).

2. W. H. Lindenberger, et al., *U.S. Patent 4,396,405* (1983).

3. W. H. Lindenberger, et al., *U.S. Patent 4,563,199* (1986).

Novel Pollution Control Methods
Group Discussion

Donald B. Wood and David C. Naugle, Shell Chemical Co., rapporteurs

Both groups discussed the process that Nalco Chemical has developed for the removal of volatile organics used in automobile paints from the exhaust air found in automobile paint spray booths. OSHA has reduced the amount of volatile organic carbon (VOC) which the automobile companies are allowed to emit to the atmosphere from their paint booths. The OSHA standard has been set based upon water-based paints under the assumption that water-based paint offers the auto manufacturers a viable paint alternative and therefore represents "best available technology". However, it is impossible at this time to obtain the high quality auto finishes customers want with these water-borne paints. The auto industry has been forced to look at expensive alternatives such as VOC control with activated carbon scrubbing while using conventional solvent-based paints.

Based upon preliminary experiments involving the removal of VOC from air with oil-in-water emulsions, Nalco was encouraged that an emulsion system could potentially solve the VOC problem and fit neatly into the systems the automakers were already using to capture paint solids in the spray booths. After presenting initial findings to the automakers, they were encouraged to

fully develop a system which could be used on a commercial scale.
In a period of less than a year Nalco moved from the bench experimental scale through a fully piloted process ready for scale up in
an automobile plant. Thus Nalco, in collaboration with an engineering firm, is able to offer a packaged solution to the automakers to solve their paint booth emission problems. This "total
package" approach to problem solving through the joint effort of
diverse companies seems to be a key method in future marketing.

Briefly the process consists of an oil-in-water emulsion
consisting of from 5-30% oil in the water phase. Paint solids,
or "sludge", is collected from the paint overspray in the water
phase and is detackified by special additives. Volatile organic
carbon dissolves in the oil phase and is later recovered by flash-
ing off the solvent after the emulsion is broken and the water and
oil phases separated. Sludge is skimmed from the water and can
economically be burned as industrial fuel. Lowering the pH of the
emulsion is used to break the oil/water emulsion into its separate
phases. The selection of the right surfactant was crucial since
the emulsion must be easily broken and then reformed without
having to add much make-up surfactant, keeping costs down. The
cost of this system is an order of magnitude cheaper than activated
charcoal for VOC recovery.

Why are anionics, such as carboxylates, sulfonates, used in
the system instead of nonionic surfactants? The emulsion must
be created at pH 8-9 to scavenge the materials and then broken
at pH 5 for proper separation and discard or recycle. However,
nonionic surfactants would be less sensitive to ions and water
hardness. If one could devise a nonionic emulsion system that
could be maintained (thermally stable) while stripping solvent,
such a system would be very desirable.

What other uses does the VOC recovery sytem have besides
spray paint? There are many vapor scrubbing applications, such
as printing processes, chemical processes.

Could such a system be used to alleviate environmental concerns about certain solvents? Yes. The oil phase of the system

can be changed to absorb specific solvents, such as chlorinated
hydrocarbons. Odor problems in the environment can be eliminated
(demonstrated with methyl ethyl ketone).

Even in this VOC recovery system, isn't there waste produced?
Yes. Sludge is produced, which contains pigments, other inorgan-
ics. But this sludge can be burned in certain approved incinera-
tors. The process does reduce the volume of sludge.

Are there flotation techniques for removing heavy metals for
disposal or separating materials for recycle? Incineration and
landfill are limited, so all waste will have to go through treatment processes in the future. Yes, there are induced air and
dissolved air techniques for extraction and washing. There are
not a lot of good technical chemical answers at present, so there
is much need for development.

Index